Validated Cleaning Technologies for Pharmaceutical Manufacturing

Destin A. LeBlanc

CRC Press
Taylor & Francis Group
Boca Raton London New York

CRC Press is an imprint of the
Taylor & Francis Group, an **informa** business

CRC Press
Taylor & Francis Group
6000 Broken Sound Parkway NW, Suite 300
Boca Raton, FL 33487-2742

First issued in paperback 2019

© 2010 by Taylor & Francis Group, LLC
CRC Press is an imprint of Taylor & Francis Group, an Informa business

No claim to original U.S. Government works

ISBN-13: 978-1-57491-116-9 (hbk)
ISBN-13: 978-0-367-39887-3 (pbk)

A CIP record for this book is available from the British Library.

Library of Congress Cataloging-in-Publication Data available on application

Visit the Taylor & Francis Web site at
http://www.taylorandfrancis.com

and the CRC Press Web site at
http://www.crcpress.com

Contents

Atomic Absorption
Ion Chromatography
Ultraviolet Spectroscopy
Enzyme-Linked Immunosorbent Assay (ELISA)
Titrations
Conductance
pH

Preface

This text is based on my experience with validated cleaning applications in pharmaceutical process manufacturing for the past 10 years. It is partly based on seminars on cleaning validation I have presented over the past 5 years to industry and regulatory groups. It is also partly based on articles and papers I have written, both journal articles and Technical Tips generated by Calgon Vestal/STERIS on various topics related to cleaning validation. Finally, it is based on the experience of the technical support group at STERIS Corporation (and Calgon Vestal) in working with pharmaceutical customers in validated process cleaning applications.

This book is designed for those who face the difficult task of designing validatable cleaning processes, and then validating those processes, in pharmaceutical process manufacturing settings. It is designed to be used by those people involved in the overall validation program. The attempt is to bring many topics together so that the interdependence of different technical areas and disciplines is evident. Seeing how each piece fits into the overall program can be valuable for streamlining the overall validation program, as well as for making each part stronger and more defendable in both internal quality audits and external regulatory audits.

It is designed to be comprehensive, covering aspects from designing the cleaning process, establishing residue limits, and selecting appropriate sampling and analytical procedures to revalidation issues. Most of the examples given in this book involve simple cleaning systems, such as the cleaning of a process vessel. The cleaning of more complex systems, such as ultrafiltration membranes or chromatographic columns, will generally require more consideration of the details, limitations, and interactions of various issues involved. Although no volume can cover all of the questions validating cleaning processes, it is hoped that this book will provide a framework for how questions can be addressed in more complex systems and for developing scientifically justified answers.

This book can be read by those entirely new to the field of cleaning validation. However, its best use is as a tool for those who have been (or are) in the trenches, doing the hard work of fitting all the pieces together. It is hoped that insight for designing a better program will be provided. For those who read this book as a first introduction to cleaning validation, it would be helpful to reread the book (or at least selected chapters) after becoming more involved in the actual work of cleaning validation. Learning can be so much more effective when combined with hands-on experience.

This book starts with cleaning objectives in Chapter 1 and moves to the cleaning process in Chapters

2–7. Issues related to validation, such as residue limits, sampling, analysis, and change control, are covered in Chapters 8–12. Chapter 13 covers special topics and/or issues for dealing with validated cleaning in different areas of the pharmaceutical industry. The last chapter covers U.S. Food and Drug Administration (FDA) expectations for cleaning validation. Each chapter has its own references for additional reading. Two appendices, one the FDA's guidance document on cleaning validation and the other a cleaning validation glossary, close out this book.

Because of the interdependence of subjects, there is some overlap and repetition between different chapters. Despite modern teaching techniques, I believe repetition is good for learning, so I make no apologies for the repetition. In addition, the repetition allows each chapter to more or less stand on its own as a valuable future reference, so that a specific topic can be researched without rereading the entire book.

In discussing regulatory requirements, I make reference mainly to the U.S. Food and Drug Administration (FDA). The FDA is not any smarter or wiser than other regulatory agencies in other countries, but they have taken the lead in establishing what is acceptable for cleaning validation. At the present time, because of the FDA's leadership, it is believed that meeting FDA expectations for cleaning validation would be acceptable in most countries.

Finally, this volume is not designed as a "how to" cookbook for cleaning validation. It does not cover such subjects as how to write a good cleaning SOP or how to write a good cleaning validation protocol. However, applying the principles covered in this book, along with the experience and documentation system in an individual facility, can help unify and simplify the process.

All of the opinions in this book are mine and do not necessarily reflect the views of my current employer, STERIS Corporation. I welcome corrections on facts or opinions in this book. Most of all, I welcome comments or suggestions you might have on topics that require further elaboration or topics that should be added for balance or completeness. Comments should be directed to me at either destin_leblanc@steris.com or destinleb@aol.com.

ACKNOWLEDGMENTS

I would like to thank both STERIS Corporation and my wife Cathy for the time they allowed me to write this book. I would also like to acknowledge the team initially at Calgon Vestal Laboratories, and now at STERIS Corporation, who had the vision and follow-through to focus on validated cleaning applications in the pharmaceutical market as a business decision. Without that business decision, I would not have had the opportunity nor the experience to write this book. I also thank the attendees at various cleaning validation seminars I have given over the past five years; I have learned as much (or more) from you as you have learned from me. Thanks should also be given to *Pharmaceutical Technology* for their kind permission to use parts of two journal articles I wrote for inclusion in Chapters 8 and 10.

I specifically thank Elaine Kopis and George Verghese for reviewing and commenting on this entire volume and Herb Kaiser, Cy Mahnke, and Sadiq Shah for reviewing selected chapters. Most of all, I thank my colleagues in the Technical Service Group in St. Louis, now called the Cleaning & Microbial Control

Technology Center, for their ideas and critical thinking that helped shape this book. You are the best!

Destin A. LeBlanc
St. Louis, Missouri
January 2000

AUTHOR BIOGRAPHY

Mr. LeBlanc is vice president of Technical Support for the Scientific Division of STERIS Corporation. He has been with STERIS for over 19 years, primarily in product development and technical service for cleaning and antimicrobial applications in manufacturing and healthcare settings. Mr. LeBlanc has 25 years, experience in technologies of specialty chemicals. He is a graduate of the University of Michigan and the University of Iowa. He had 9 patents and over 25 publications in his fields of expertise. Mr. LeBlanc has lectured on issues related to contamination control in pharmaceutical manufacturing facilities in North America, Europe, Asia, and Australia. He regularly trains FDA (U.S. Food and Drug Administration) investigators on cleaning validation issues. He is a member of PDA (Parental Drug Association) and ISPE (International Society of Pharmaceutical Engineering) and is on the faculty of the PDA Training and Research Institute.

Mr. LeBlanc can be reached as follows:
  7405 Page Avenue
  St. Louis, MO 63133 U.S.A.
  314-290-4790
  314-290-4650 (fax)
  destin_leblanc@steris.com

1

Cleaning Objectives

In doing any task, in addition to knowing what is to be accomplished, it helps to also know why it is being done and what limitations may be placed on the process for doing it. With that information, one can certainly devise a better (more efficient) way to accomplish the task. This principle also applies to the cleaning of process equipment in a pharmaceutical manufacturing facility. The goal is to have an acceptable cleaning process. The first question that can be asked is, "Why do pharmaceutical facilities want to clean?" The answers that may be given are generally one or more of the following:

- To protect product integrity

- To reuse the equipment

- Because regulatory authorities require it

Each of these topics will be covered below. The regulatory background will be covered in more depth,

1

particularly the recent history, since this can help one understand why the state of cleaning validation has been somewhat in flux.

PRODUCT INTEGRITY

Maintaining product integrity first includes preventing cross-contamination, in which one drug active from the product just cleaned becomes an unacceptable contaminant in the next drug product (with a different active) manufactured in the cleaned equipment. Several issues arise here. First is the possible pharmacological effect of the contaminating residue in the subsequently manufactured product. This is generally the major concern and is the usual basis for calculating residue acceptance limits for cleaning validation protocols. Another concern from the contaminating residue of a drug active is possible drug interactions between the contaminating drug active and the intended drug active in the contaminated product. This drug interaction may result in an adverse pharmacological effect, in a reduced pharmacological effect of the drug active due to effects on bioavailability by interaction with the residue of the contaminating active, or in reduced shelf life and/or instability due to the interaction. These interaction effects may be more difficult to assess. If they are theoretical possibilities, the actual effects can often be verified by experiments, for example, spiking an active with known amounts of potential contaminants to determine changes in bioavailability.

Another issue in product integrity involves contamination not necessarily with another drug active but with drug excipients, cleaning agents, and/or equipment residues (such as rouge and particulates from equipment wear). Such residues do not have pharmacological effects per se. However, they may result in effects such as changes in the bioavailability or the stability of the contaminated drug products. Residues from cleaning agents and from equipment may also pose special safety concerns. Particulates from rouge or worn gaskets may pose an unusual risk in parenteral products.

A third issue in product integrity involves microbial and/or endotoxin contamination. Clearly, this can present safety concerns, especially with parenteral products. These concerns may also arise with oral and dermatological preparations, depending on the type and level of contaminating species. An additional concern from microbiological contamination involves effects on the stability or shelf life of the finished drug product. Microbial contamination is more difficult to assess because, unlike chemical residues, which may be relatively unchanged as they are transferred from cleaned equipment surfaces to subsequently manufactured products, microorganisms may possibly rapidly proliferate in a drug product. It should also be noted that, depending on the nature of that drug product, microorganisms may be rapidly eradicated.

A special case of product integrity involves maintaining lot integrity on a dedicated product line or in a product campaign. Cleaning between lots of the same product is done for several reasons. One reason is that it may be required for proper equipment function. For example, the buildup of residues may interfere with proper tablet formation. A second reason is to maintain lot integrity. If residues of the previous lot are not adequately removed, it may be difficult to maintain lot or batch integrity. Failure to maintain lot integrity may be a significant issue if any one lot in a campaign is involved in a potential recall.

EQUIPMENT REUSE

If all pharmaceutical manufacturing equipment were disposable, cleaning and cleaning validation would be of little concern. Imagine a future scenario in which all manufacturing equipment is disposable: After manufacturing a pharmaceutical product, the equipment is crushed and sent to an incinerator for disposal. For most equipment today, this ideal situation (at least from the cleaning validation point of view) is rarely the case. Most manufacturing equipment is stainless steel or glass lined and is relatively expensive. High capital costs require that the equipment be reused. Therefore, such equipment should be adequately cleaned (at least the product contact surfaces) in a validated process. One should be aware that there may be parts of the equipment system that may be effectively viewed as "disposable." On a case-by-case basis, the costs and risks of cleaning (and validating that cleaning) should be weighed against the costs and risks of using a new piece of equipment and disposing of it after use. Examples of equipment that may be considered disposable include silicone tubing and plastic scoops.

REGULATORY REQUIREMENTS

The focus here will be on the U.S. Food and Drug Administration (FDA). The FDA has taken the lead in requiring cleaning validation and in helping shape what is expected in a cleaning validation program. In this time of harmonization, regulatory agencies from other countries are looking to the FDA for leadership, and pharmaceutical companies with significant international business are looking to meet the expectations of the FDA [1,2].

Cleaning is not something that just appeared in the late 1980s. Pharmaceutical companies have always practiced cleaning, and cleaning has always been a part of Good Manufacturing Practices (GMPs). A review of the cGMPs (21 CFR Parts 210–211) shows many statements, paraphrased below, that are related to the cleaning process [3,4].

- *211.42:* Buildings shall be of suitable size, construction, and location to facilitate cleaning.

- *211.42(c)(10)(v):* A system shall be maintained for cleaning and disinfecting the aseptic processing room and equipment to produce aseptic conditions.

- *211.56(a):* Buildings shall be maintained in a clean and sanitary condition.

- *211.56(b):* There shall be written procedures assigning responsibility for sanitation and describing in sufficient detail the cleaning schedules, methods, equipment, and materials to be used in cleaning.

- *211.56(c):* There shall be written procedures for the use of cleaning and sanitizing agents.

- *211.63:* Equipment shall be of the appropriate design for its cleaning.

- *211.67(a):* Equipment and utensils shall be cleaned, maintained, and sanitized at appropriate intervals.

- *211.67(b):* Written procedures shall be established and followed for the cleaning of equipment. These procedures shall include assignment of responsibility for cleaning, maintenance of cleaning and sanitizing schedules, a sufficiently detailed description of methods used for cleaning (including disassembly as well as assembly, protection of clean equipment from recontamination, and inspection of equipment for cleanliness immediately before use).

- *211.94:* Containers and closures shall be clean and, where appropriate, sterilized and depyrogenated. Methods of cleaning, sterilizing, and depyrogenating shall be written and followed.

- *211.105(a):* All processing lines and major equipment shall be properly identified to indicate their contents and, when appropriate, the phase of processing.

- *211.111:* Time limits for the completion of each phase of production shall be established.

- *211.113:* Appropriate written procedures to prevent microbiological contamination shall be established and followed.

- *211.182:* A written record of major equipment cleaning shall be included in equipment logs.

- *211.188:* Records shall include documentation that each significant step in the manufacture was accomplished.

Clearly, these GMPs require that cleaning SOPs (Standard Operating Procedures) be in place and the cleaning processes be documented. What is new is that certain cleaning processes must be validated.

This requires some higher assurances of consistency and control in the cleaning process. Cleaning validation became a "hot" issue around 1990. One key case, involving contamination of a drug product by a pesticide residue (the equipment for the bulk active drug was cleaned with solvent reclaimed from pesticide manufacture), was attributed to a poorly controlled cleaning process [1].

The Barr Labs decision [5] is regarded as a critical case acknowledging the FDA's right to require that cleaning processes be validated. In that case, the FDA had identified problems at Barr Laboratories related (among other things) to Barr's cleaning practices. The FDA requested that Barr validate its cleaning procedures. Barr objected that validation of cleaning was not required by the cGMPs but still proceeded with a cleaning validation program. However, the FDA was dissatisfied with the extent of Barr's cleaning validation. In the *U.S. v. Barr Laboratories* decision, the right of the FDA to require cleaning validation was upheld. The court also agreed that cleaning validation was not limited just to "major equipment"; companies also have to adequately describe the cleaning agents used. On the other hand, the court held that testing for residues of cleaning agents was not necessarily required, and one successful cleaning procedure may be "not insufficient" for cleaning validation. Clearly, the force of these last two points was lost in subsequent activities, since cleaning agent residues are generally of significant concern for the FDA, and the rule of three Process Qualification (PQ) runs for process validation generally applies to cleaning validation.

At the time this case was being prosecuted, the FDA had issued its *Biotechnology Inspection Guide,* which called for the "validation of cleaning procedures for the processing of equipment," including the comment that this was "especially critical for a multiproduct facility." It was in this guide that the FDA first stated that residue limits must be "practical, achievable, and verifiable" [6]. In July 1992, the Mid-Atlantic Region of the FDA published the *Mid-Atlantic Region Inspection Guide: Cleaning Validation* [7]. This document covered in more detail the expectations for cleaning validation, including equipment design, SOPs and documentation, analytical methods, sampling procedures, limits, and detergents. This document was revised with a new introduction and minor wording changes in May 1993 [8]. Neither document was an official FDA guidance document, although a foreword to the later document by the regional FDA director clearly stated the expectation that the Mid-Atlantic Region would use the document in its inspections.

In July 1993, the official guidance document, *Guide to Inspections of Validation of Cleaning Processes,* was issued [9]. It followed the same major topics as the earlier Mid-Atlantic guide but included significant changes in wording as related to the various topics. More details of the expectations in this document will be discussed later in Chapter 14. For now, the significant item is that the FDA clearly stated that cleaning processes should be validated *and* also gave specific guidance on expectations involving some elements of SOPs and validation protocols.

The next major regulatory step was the proposed revision of the GMPs in May 1996 [10]. These proposed amendments (Sec. 211.220) require that "the manufacturer shall validate all drug product manufacturing . . . steps in the creation of the finished product including . . . cleaning." This is significant in that the requirement for cleaning validation, if the amendments are approved, will be clearly written in the GMPs. This is no real change for the FDA, because they have already been enforcing cleaning validation as if it were a clear mandate in the GMPs. A good case can be made for their approach. Cleaning is clearly required by the GMPs. Cleaning can be a critical process for drug manufacture. Critical process steps in the manufacture of drug products should be validated. Therefore, cleaning should be validated. It should also be pointed out that even though these GMP amendments address finished drug products, the FDA is currently taking the same approach to cleaning validation in the manufacture of bulk actives or active pharmaceutical ingredients (APIs). It is useful to recall here that some of the first cases that raised the awareness of the need for cleaning validation involved the cleaning not of finished drug products but rather of bulk actives. The current state of the industry and regulatory climate is that pharmaceutical companies are without excuse if they don't think the FDA is serious about cleaning validation for critical cleaning activities.

OTHER OBJECTIVES

Before the advent of the validation of cleaning processes, pharmaceutical companies had the flexibility to change cleaning procedures as needed. When a process meets the rigors and change control elements of process validation, this flexibility is lost. Of course, what is lost in flexibility is made up in consistency and control of the cleaning process. However, because most manufacturers want to avoid changes once a process is validated, there is an increasing awareness of the need to select a cleaning process that can meet the company's needs not only in the present but also in the future. Therefore, needs for validation should be balanced with other business considerations in selecting cleaning processes. For example, some pharmaceutical companies would purchase cleaning agents that were either retail products, "janitor supply" institutional products, or cleaning agents designed for other cleaning applications in other industrial applications. With such products, the suppliers may frequently offer "new and improved" versions of the same product or may discontinue products based on the needs of other industries that "drive" the need for that product. If consistency (consistent formulation and availability) is a requirement for validated cleaning, the types of cleaning agents mentioned above may not be the best choices, since significant change control and/or revalidation activities may be required if and when the formulation or availability of such products is changed.

There are a variety of other objectives that may impact the choice of the cleaning process. Topics to be discussed here include the following: solvent reduction, shorter cleaning times, increased equipment utilization, extension of equipment life, multiproduct facilities, worker safety, and cost-effectiveness.

Solvent Reduction

Solvents such as acetone and hexane are commonly used in cleaning operations in bulk pharmaceutical manufacture. They are used because the solvent is already used in the manufacturing process and because the bulk active may be readily soluble in the solvent. However, with increasing concern over solvent emissions, many major pharmaceutical companies have a corporate goal of reducing solvent use and solvent emissions [11]. A switch to aqueous-based cleaning procedures, which may require significant reengineering and cycle development, may both reduce the use of solvents and produce a more consistent, lower cost cleaning process.

Shorter Cleaning Times

A second objective is that of obtaining a quicker turnaround of equipment from batch to batch or from product to product in a manufacturing setting. If companies may no longer use repeated cleaning processes (i.e., wash until clean) to assure that equipment is acceptably clean, then a fixed cleaning procedure, with a more or less fixed cleaning time, is required. Ideally, the cleaning time should be as short as possible, within the constraint of still being a robust cleaning procedure. Shortening a 3-hour cleaning procedure by 15 minutes may not be significant; however, converting from a 2-day cleaning process to a 10-hour cleaning process may be significant.

Increased Equipment Utilization

The drive to shorter turnaround times is tied to increased equipment utilization. New equipment is often purchased to give greater flexibility in the manufacture of products. This sometimes means it is more complex in design, which may require more elaborate cleaning procedures, or at least more cycle development work to arrive at a validatable cleaning process.

Extension of Equipment Life

With increased equipment utilization, the issue of deleterious effects of the cleaning process on the cleaned equipment is of more concern. With glass-lined vessels, the major issue relating to deleterious effects is the use of highly caustic solutions for extended time periods at higher temperatures. Such use can lead to etching of the glass surface [12]. Such an etched surface may, in addition to affecting efficient processing, result in a surface that is much more difficult to clean. This may eventually result in a need for revalidation of the cleaning process. With stainless steel vessels, there are two concerns. One is the leaving of residues on surfaces, which can result in underdeposit corrosion. At a minimum, a validated cleaning process should result in no visible residues left on surfaces; therefore, this type of corrosion should not be an issue. In fact, the best way to maintain a passivated surface on stainless steel is to keep the surface clean and free of deposits and/or residues. A second concern with stainless steel is the extended use of hypochlorite-containing cleaners, which has been known to cause corrosion and rouging of stainless steel surfaces. Because of these various concerns, more attention needs to be paid up-front to selecting a cleaning process that minimizes deleterious effects, thus extending equipment life and minimizing special maintenance procedures.

Multiproduct Equipment

The issues of shorter cleaning times, increased equipment utilization, and extending equipment life are also related to the trend toward multiproduct process trains and facilities. Cleaning and cleaning validation would be simplified if equipment and/or facilities were dedicated to one product. In many cases, this is not economically feasible. The cleaning needs of multiproduct equipment are much more complex and are complicated by the fact that cleaning validation is concerned not only with the cleaning and removal of the drug just manufactured but also with setting residue acceptance limits based on the possible contamination of the *next* drug produced in that same equipment. Acceptable cleaning procedures, therefore, always depend to a given extent on what other products are manufactured on the same equipment. In addition, if a new product is planned for introduction into manufacturing equipment previously validated with a cleaning process, the effect of the new product on the previously completed cleaning validation work, and particularly on residue acceptance limits, should be evaluated. Another factor that impacts multiproduct facilities is the possibility of grouping products together and using one cleaning process for all products in that group. If a "worst-case" product can be established, then cleaning validation can be performed on that worst case to represent the entire group, and all products within that group can be covered by one cleaning validation protocol [13].

Worker Safety

While control and consistency have been one reason for companies to switch from manual to automated cleaning, a second reason has been increased concerns over worker safety. Many aqueous cleaning agents are either high or low pH solutions and can cause significant tissue damage in contact with the skin or eyes. Automated processes certainly present the opportunity to minimize worker exposure to cleaning solutions.

Cost-Effectiveness

Many of the previous objectives are also tied to the concept of cost-effectiveness. While restraints on selling prices of drugs have always been an issue outside the United States, market pressures restraining price increases within the United States have become a reality in the 1990s. Increased costs of manufacturing can no longer be automatically passed on to insurers and consumers. Therefore, the goal of many large manufacturers is low costs, since controlling costs is one way to increase profitability. While controlling the cost of cleaning is a valid objective, a requirement is no sacrifice in quality of the cleaning process. In fact, quality is often significantly improved because manufacturers are paying closer attention to the issues involved in achieving a validated cleaning process.

The driving force for cleaning validation has been regulatory pressure. However, this has just been a mandate that cleaning validation be done. The exact way that it is performed in different facilities has been shaped to a large extent by some of these factors discussed above.

ASSURANCE OF CLEANING

Both cleaning validation and cleaning verification are methods of showing that the cleaning of process equipment is performed adequately so as not to affect the safety or efficacy of the next drug product manufactured in the cleaned equipment. *Validation* is usually defined as documented evidence with a high degree of assurance that a process consistently meets its predetermined quality attributes. As applied to a cleaning process, this means evaluating a cleaning SOP on specified equipment after the manufacture of certain drug products (or APIs). It involves setting acceptance criteria (visually clean and specific analytical limits based on possible contamination of subsequently manufactured products). It also involves developing data on three consecutive cleaning runs. If successfully validated, the cleaning process is then maintained under a change control program.

Cleaning validation is applicable to processes that are done repeatedly. (This includes a consistent cleaning SOP and a consistent soil/drug on the equipment surfaces to be cleaned.) However, there are cases where the cleanliness of the equipment must be assured, but the cleaning process is not done repeatedly (or frequently enough) to conduct three PQ runs. Examples include clinical trial materials and products manufactured only once a year. With drug products manufactured only once a year, it is probably not reasonable to expect that something in the cleaning process (nature of soil, equipment, etc.) would stay the same (for validation purposes) over the two years necessary to obtain three PQ runs. With most clinical trial materials, batch sizes and equipment may vary, process parameters may be "tweaked," and other changes may be made such that three consecutive cleaning runs with the same process may be highly unlikely (even though the three runs may be performed over a short period of time).

In cases such as these, the FDA expects that the cleaning process be verified [14]. Cleaning *verification* is similar to cleaning validation in that it involves documented evidence as well as a high degree of assurance. However, it differs from validation in that the data generated apply only to that specific cleaning event. Since the process is assumed to possibly change, the data cannot be assumed to apply to other similar cleaning processes. (Although the data would be suggestive about other similar cleaning events, verification fails the "consistency" requirement for validation purposes.) Also, while specific cleaning analyses are performed, the acceptance criteria may not be predetermined. Acceptance criteria depend on the subsequently manufactured products; in a clinical trial setting, the next product may not be known at the time of cleaning. Note, however, that in this case an evaluation should be done to determine whether the residues actually found in the cleaned equipment would be acceptable once the next product to be manufactured is selected (based on that next product's dosing and batch size).

Many of the same things done in validation are also done in verification. For example, in verification, there is a cleaning SOP, and there are techniques (visual and analytical) to determine the levels of cleanliness. This may include various swab and rinse samples. In fact, with verification, there may be more analytical work (both in terms of the number of sampling points and analytical tests performed) done. Why more? Because verification deals with specific cleaning events, whereas validation involves a repeating, consistent cleaning process. In validation, there is considerable prequalification work to address items such as worst-case locations. In verification, by contrast, this prequalification work may not be possible or may be done to a more limited extent. Therefore, more sampling locations and possibly more analytical procedures may have to be performed as compared to what is done in a validation mode.

In addition, while cleaning validation addresses the universe of possible products that may be subsequently manufactured following the cleaning process, in cleaning verification the next product may not be known at the time cleaning is performed. Therefore, prior to the next manufacturing process in the cleaned equipment, it is necessary to calculate and determine whether the residues actually found are acceptable considering the dosing and batch size of the next product. It is at least conceivable that cleaning may have to be repeated (and verified) depending on the results of such calculations.

Cleaning verification may also be used in cleaning after an invasive maintenance procedure on equipment that is under change control, to confirm that the equipment is still acceptably clean for use. Note that the testing done after such maintenance may be different from the testing done in the original cleaning validation protocol because the nature and location of residues may be different. Cleaning verification may also be utilized if cleaning is to be repeated due to a deviation in the cleaning process. For example, if the cleaning SOP calls for the cleaning process to begin within 12 hours after the end of the manufacturing process, and if the equipment (for whatever reason) cannot be cleaned until 24 hours after the end of processing, then the manufacturer may verify cleaning following the delayed cleaning procedure.

REFERENCES

1. FDA. 1993. *Guide to inspections of validation of cleaning processes.* Rockville, Md., USA: Food and Drug Administration, Office of Regulatory Affairs.

2. PIC. 1999. *Recommendations on cleaning validation.* Document PR 1/99–1. Geneva, Switzerland: Pharmaceutical Inspection Convention.

3. 21 CFR 210: *Current good manufacturing practice in manufacturing, processing, packing, or holding of drugs; General.* 1 April 1997 (revised).

4. 21 CFR 211: *Current good manufacturing practice for finished pharmaceuticals.* 1 April 1997 (revised).

5. *United States v. Barr Laboratories*, 812F. Supp. 458 (DNJ 1993).

6. FDA. 1991. *Biotechnology inspection guide: Reference materials and training aids.* Rockville, Md., USA: Food and Drug Administration, Office of Regulatory Affairs.

7. FDA. 1992. *Inspection guide: Cleaning validation.* Rockville, Md., USA: Food and Drug Administration, Mid-Atlantic Region.

8. FDA. 1993. *Inspection guide: Cleaning validation,* rev. Rockville, Md., USA: Food and Drug Administration, Mid-Atlantic Region.

9. FDA. 1993. *Guide to inspections of validation of cleaning processes.* Rockville, Md., USA: Food and Drug Administration, Office of Regulatory Affairs.

10. FDA. 1996. Current good manufacturing practice; Proposed amendment of certain requirements for finished pharmaceuticals. *Federal Register* 61:20103.

11. EPA. 1998. *Fact sheet: Final air toxics rule for pharmaceutical production.* Washington, D.C.: Environmental Protection Agency, Office of Air and Radiation.

12. 3009 Glass. Brochure 3009 USA 1/15/92. Union, N.J., USA: DeDietrich, Inc.

13. PDA. 1998. *Points to consider for cleaning validation.* PDA Technical Report No. 29. Bethesda, Md., USA: Parenteral Drug Association.

14. FDA. 1997. *Human drug cGMP Notes,* Vol. 3, No. 5. Rockville, Md., USA: Food and Drug Administration, Center for Drug Evaluation and Research.

2

Cleaning and Cleaning Agents

There are two main aspects of cleaning in any application. One is the chemistry of cleaning, which will be discussed in this chapter. The other involves the engineering aspects of cleaning, including the cleaning method and various process parameters; these will be covered in Chapters 3, 4, and 5. The topics covered in this chapter include cleaning mechanisms, cleaning agent options, aqueous cleaning options (since the trend in general has been toward aqueous cleaning), and combination cleaning processes.

CLEANING MECHANISMS

Cleaning involves removing an unwanted substance (the contaminant) from a surface (the equipment to be cleaned). The chemistry of cleaning includes several mechanisms that serve to remove or assist in removing the contaminants from the equipment surfaces [1]. Understanding (or at least being aware of) cleaning mechanisms can assist in the selection of the proper cleaning agent; more importantly, it can assist in the proper design of the overall cleaning process. The cleaning mechanisms, with their features and limitations, covered in this chapter include the following:

- Solubility

- Solubilization

- Emulsification

- Dispersion

- Wetting

- Hydrolysis

- Oxidation

- Physical removal

- Antimicrobial action

Solubility

Solubility involves the dissolution of one chemical (the contaminant) in a liquid solvent [2]. For example, salts may be soluble in water, and certain organic actives may be soluble in acetone or methanol. This is one of the primary cleaning mechanisms and, other things being equal, is a preferred mechanism because of its simplicity. Unfortunately, there are several complicating factors. One factor is that information about the solubility of a compound in a solvent does not necessarily address the rate of solubility. This can be illustrated with the dissolution of sugar in water. While sugar is soluble in water, placing a spoonful of granulated sugar in a glass of water will not necessarily result in the rapid or immediate dissolution of the sugar. In many cases, even after an hour, the sugar will still be evident on the bottom of the glass. In order to make the sugar dissolve more readily, one must either stir the glass of water (providing agitation) or heat the water so it is warm (or both). The same may be true with soluble contaminants or residues in pharmaceutical cleaning. In addition, the physical form of the contaminant may affect dissolution rates. Continuing with the example of sugar, dissolution rates with granular sugar may be relatively rapid (with appropriate agitation and/or heat). However, if the sugar exists in the form of a hard candy (such as so-called "rock" candy), dissolution rates may be significantly slower, even with the use of agitation and heat.

A second issue involving solubility is whether the contaminant, as it exists in the equipment to be cleaned, has been significantly altered such that it is no longer readily soluble. For example, the heat used in pharmaceutical processing may chemically alter the contaminating residue such that it is no longer water soluble. Continuing with the example of sugar, the sugar may become "caramelized" (polymerized) during processing and form a relatively water-insoluble material on equipment surfaces. Using water to clean the equipment, even with adequate agitation at elevated temperatures, may no longer serve to adequately remove the sugar residues.

A third issue in solubility applies to cases where there is more than one chemical species in a residue to be cleaned. This is most common in finished pharmaceutical manufacturing, in which residues will typically contain both the drug active as well as a variety of excipients. In these cases, it is possible that the solubility profile of the active may be considerably different from the solubility profile of the excipients. This is most evident in tablets, where the water solubility of the excipients and/or coating materials is usually much less than the water solubility of the actives. In designing a cleaning solution that will dissolve the residue, which residue is targeted? Unless all chemical species are clearly soluble in the selected solvent (water or an organic solvent), solubility may not be the best cleaning mechanism to select if it is the sole cleaning mechanism.

Solubilization

Solubilization is similar to solubility, except that it involves an additive to the pure solvent to render the residue soluble [3]. If the solvent is water, this usually involves the addition of a surfactant, a pH modifier, or a water-miscible organic solvent to solubilize the residue. For example, modifying of the solvent water by adding of potassium hydroxide to a pH of 12 will result in certain organic compounds, which are normally water insoluble, becoming water soluble. Stearic acid is water insoluble. However, a cleaning solution of potassium hydroxide in water will solubilize the stearic acid (effectively converting it to potassium stearate). One alternative way of viewing this specific situation is that one is chemically modifying the residue to convert it from an acid to a potassium salt, which is readily water soluble. This mechanism may apply to a variety of organic species with carboxylate groups.

The use of certain surfactants or water miscible solvents (such as glycol ethers) may have similar functional effects in terms of solubilizing the residue. It may be important to understand this as a cleaning mechanism in an individual situation because of the limits this puts on the process. For example, if the cleaning process involves solubilization of a carboxylate by converting it to a salt, then it is not simply a matter of having water at pH 12. The alkalinity source and amount are important. For example, in the case of stearic acid cited previously, using sodium hydroxide at pH 12 would not be adequate because sodium stearate is water insoluble, whereas potassium stearate is water soluble. In addition, the total amount of potassium hydroxide should be at least a stoichiometric amount needed to convert the acid to a salt.

This mechanism also impacts neutralization of the spent cleaning solution. A common question is whether the spent cleaning solution can be neutralized in situ before it is discharged to the drain. This is a bad practice in general. If solubilization is the cleaning mechanism employed, then a change in pH, for example, from pH 12 back to a neutral pH of 8, may result in a solubilized residue coming back out of solution and redepositing on equipment surfaces. Any neutralization of such a cleaning solution should be done outside the equipment to be cleaned, preferably in a separate holding tank.

Emulsification

For cleaning purposes, emulsification is the process of "breaking up" an insoluble liquid residue into smaller droplets and then suspending those droplets throughout the water [4]. The breaking up process is usually accomplished by applying mechanical energy to the system. The emulsion is usually stabilized by the addition of surfactants or polymers. Milk is a common example of an emulsion; it contains fat (the cream) emulsified in water. In cleaning processes, examples of emulsions formed as a result of cleaning include aqueous emulsions of mineral oil, silicone oils, or petrolatum. Mechanical energy is supplied in the form of agitation or turbulence. Emulsions are stabilized usually with the addition of anionic or nonionic surfactants to the water.

Emulsions are thermodynamically unstable. At some point in time they will separate, with the insoluble liquid residue usually floating to the top (or to the bottom in the case of dense insoluble phases). In the case of milk, the emulsion is very stable, and other problems (microbiological contamination) usually arise before the milk would separate in normal use. Emulsions formed in cleaning operations are generally much less stable than an emulsion like milk. Usually upon discontinuing the input of mechanical energy (discontinuing the agitation, for example), the emulsion may start to break. This may be in a very short time (say, 5 to 10 minutes), or it may be over an extended period of many hours. This breaking of the emulsion may result in the redeposition of the cleaned residue back onto the equipment surfaces, clearly an undesirable feature.

For this reason, emulsions should continue to be agitated up until (and perhaps during) the time the cleaning solution is discharged to the drain. What was discussed regarding neutralization in the process equipment in the previous section also applies here. Any change in the pH by neutralizing an emulsion may affect the quality of that emulsified residue. If pH neutralization in the process tank causes the emulsion to break, then any value added by emulsifying the residue is quickly lost by having it redeposited onto the cleaned equipment surfaces.

Dispersion

Dispersion is similar to emulsification, except that it involves the wetting and deaggregation of solid particles and then the subsequent suspension of those particles in water [5]. A common example of a dispersion is prepackaged salad dressing. This involves particles (usually spices) suspended in a water/oil emulsion. One way to think of dispersions is that it is an emulsion using solid particles instead of liquid particles. For solid dispersions, the use of surfactants (such as anionics) and mechanical energy (from the agitation of the liquid or liquid flow) are used to wet and deaggregate the solid particles. Continued mechanical energy, as well as the use of certain dispersants (usually charged, low molecular weight polymers, such as polyacrylates), is used for the suspending process. Dispersion as a process is probably more important in dry product manufacture, such as powder blending and tablet manufacture. As with emulsions, a key feature of processing is that the mechanical input of energy continue up to the time (and perhaps during the time) the cleaning solution dispersion is discharged to the drain.

Wetting

Wetting involves the displacement of one fluid (in most cases air) from a solid surface by another fluid (the cleaning solution) [6,7]. Wetting by water is improved by the addition of surfactants to lower the surface tension. Pure water has a typical surface tension of about 73 dynes/cm at 18°C. Wetting can be improved by the addition of surfactants, which can lower the surface tension to a range of about 30–40 dynes/cm. By contrast, the addition of sodium hydroxide alone to water has either no effect or slightly increases the surface tension. Wetting of surfaces involves wetting of the soil to be removed as well as wetting of the surface to be cleaned. Both are important in the cleaning process. Wetting of the soil to be removed provides for more rapid dissolution, solubilization, emulsification, and dispersion. An additional benefit of a lowered surface tension is illustrated in an exaggerated way in Figure 2.1 in terms of better penetration of the cleaning solution into cracks and crevices, which are usually difficult-to-clean locations.

Hydrolysis

Hydrolysis involves the cleavage of certain bonds in an organic molecule. This cleavage usually involves esters or amides. Hydrolysis is accomplished in

Figure 2.1. Wetting of soils and surfaces with and without surfactants. (Note: Effect is exaggerated for illustration purposes only.)

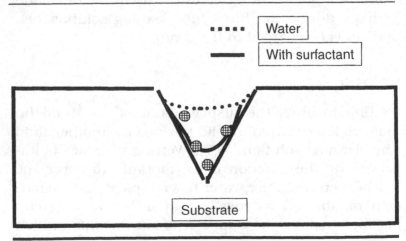

aqueous solutions, usually under alkaline or acidic conditions, and usually at elevated temperatures [8]. Figure 2.2 illustrates the hydrolysis of an ester under alkaline conditions, producing the salt of a carboxylic acid and an alcohol. The time and temperature required for hydrolysis will depend on the specific ester or amide being hydrolyzed as well as the nature and amount of the residue (because this usually involves hydrolysis of a relatively water insoluble material). Hydrolysis can be part of an effective cleaning procedure because it can convert a relatively large, water-insoluble molecule into smaller, more water-soluble molecules. Water solubility is increased in the hydrolysis process partly because the resultant molecule is smaller and partly because the resultant molecule is more polar.

It should be noted that hydrolysis by itself is not enough; the resultant hydrolyzed residues must either be water soluble or solubilized at the pH of the cleaning solution. It is entirely possible that esters may hydrolyze, but the resultant fragments may not be adequately water soluble for effective cleaning.

Figure 2.2. Hydrolysis of organic ester to produce more water soluble soils.

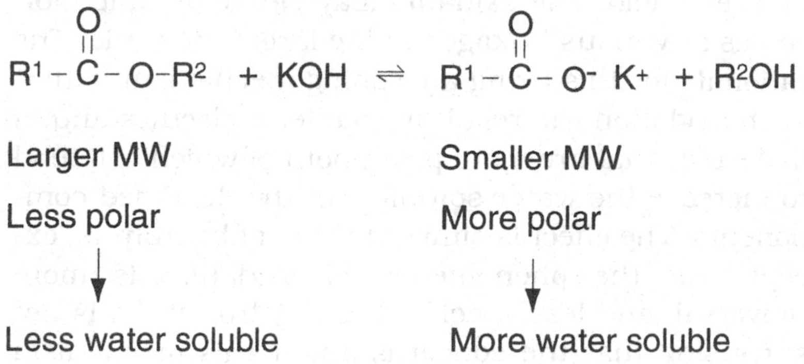

$$R^1 - \overset{\overset{\textstyle O}{\|}}{C} - O - R^2 \ + \ KOH \ \rightleftharpoons \ R^1 - \overset{\overset{\textstyle O}{\|}}{C} - O^- \ K^+ \ + \ R^2OH$$

Larger MW	Smaller MW
Less polar	More polar
↓	↓
Less water soluble	More water soluble

This has to be determined experimentally on a case-by-case basis.

This mechanism for cleaning can be important in terms of selecting the analytical methods for residue determination after cleaning. If the active agent is an ester that degrades during the cleaning process (forming a carboxylic acid salt and an alcohol), it may not be valid to target a specific analytical method for the active agent for residue determination for cleaning validation. In such a scenario, the detection of the active agent indicates that the active agent is present after cleaning. However, the absence of the active ingredient in the analyzed samples does not necessarily mean that the system is adequately cleaned, because absence of the active agent is expected just in the presence of the washing solution. The residues present (if any) would be expected to be the hydrolysis products (the carboxylic acid and the alcohol), not the ester itself.

Oxidation

Oxidation involves the cleavage of various organic bonds, such as carbon-carbon bonds, by the action of a strong oxidizing agent. Strong oxidizing agents that may be present in a cleaning situation include species such as sodium hypochlorite, hydrogen peroxide, and peracetic acid. The oxidants may cleave organic molecules at various linkages in the larger molecule. The rationale for this being a cleaning mechanism is that such oxidation will result in smaller molecules and in molecules that are more polar, both of which will tend to increase the water solubility of the degraded components. The effect is similar to that of hydrolysis, except that the phenomenon of oxidation is more universal (and less specific) than hydrolysis. It is not surprising that the concerns about this mechanism in a cleaning process are similar to the concerns with hydrolysis, namely that the resultant degradation products must be water soluble for effective cleaning, and that analytical method selection may be different because one would not expect the unoxidized residue to be present after cleaning. Therefore, it may not be appropriate to target a specific method to analyze that unoxidized residue.

Physical Cleaning

While most of this discussion has focused on the chemical mechanisms of cleaning, one simple mechanism of cleaning is physical removal by using some mechanical force. This may be hand scrubbing during a manual cleaning operation or cleaning manually with a high pressure water spray. In both of these cases, the objective is to physically dislodge the residue, where it is then carried away from the surface by the high pressure water stream or by the scrubbing action. In such cases, the cleaning may be assisted by use of a surfactant in the cleaning solution to assist in the wetting of the residue. A related form of physical cleaning is the mechanical action due to a moving stream of water (or solvent). Before introducing the cleaning solution into the equipment, it is common in pharmaceutical manufacturing to prerinse the equipment with ambient temperature water. This prerinsing serves to help physically remove gross contamination, thus leaving less of the contaminating residue for the cleaning solution to emulsify, disperse, hydrolyze, and so on. The effectiveness of physical removal will depend on the nature of how the residue is attached to the surface. For example, "baked on" residues may not be easily prerinsed from surfaces. The importance of physical processes in cleaning should not be minimized, as physical processes are to some extent involved in all cleaning mechanisms.

Antimicrobial Action

Antimicrobial action may also be considered a special type of cleaning mechanism. In discussing this as a cleaning mechanism, it is important to separate those mechanisms that may kill organisms but leave behind nonviable microbial residues (such as in steam sterilization) from those mechanisms that kill organisms but may assist in the further removal of nonviable residues (using an oxidizing biocide). This can be illustrated with the common mold found in bathrooms in humid climates. It is possible to kill the *Aspergillus niger* mold with a phenolic disinfectant. However, the resultant surface is still discolored black, even though the mold is dead. On the other hand, if the antimicrobial agent is an oxidizing biocide such as sodium hypochlorite, the mold will be killed *and* the black stain will be oxidized (bleached) to present a visibly clean surface. Further discussions of antimicrobial mechanisms are beyond the scope of this book.

Real-Life Situations

In real-world cleaning situations, cleaning may involve a variety of these mechanisms [9]. Particularly when one is dealing with finished drug products in which there are a variety of chemical types as residues, a variety of cleaning mechanisms may be involved. An alkaline, aqueous surfactant–containing cleaner may be used to emulsify an excipient, while at the same time causing the hydrolysis of the active agent so that the resulting hydrolysis products are solubilized at the higher pH. In this case, at least three of the mechanisms may be utilized. It may be very difficult to separate out which cleaning mechanisms are actually involved and the relative importance of each mechanism. This is not necessarily a problem. However, awareness of cleaning mechanisms may alert one to certain considerations in designing cleaning processes, such as the necessity for agitation until the cleaning solution is drained or the selection of an analytical procedure to target the residue that might be present (if any) after the cleaning process.

CLEANING AGENT OPTIONS

There are a variety of cleaning agent options available to pharmaceutical companies for their cleaning processes [10]. A first classification of cleaners is into organic solvents and aqueous-based cleaners. Organic solvents, including solvents such as acetone, methanol, and ethyl acetate, are most commonly used for cleaning in bulk drug manufacture. Aqueous cleaning includes the use of water alone or the use of commodity chemicals or formulated specialty cleaners diluted in water. Commodity chemicals include inorganics such as sodium hydroxide and phosphoric acid. Formulated speciality cleaners include a variety of liquid, multifunctional products containing surfactants and a variety of functional additives. These cleaners are generally liquid (as opposed to dry powder) products because of the ease of using liquids in cleaning agent feed systems.

Organic Solvents

There are a variety of advantages and disadvantages of organic solvent cleaning as opposed to aqueous cleaning, particularly in a bulk active manufacturing facility. In a bulk facility, the emphasis is on cleaning only specific compounds (and closely related analogs). Therefore, the issue of solubility being the primary cleaning mechanism is more straightforward. If the active being manufactured is known to be soluble in an organic solvent, then that solvent is an appropriate cleaning agent. Issues like agitation and temperature, however, may have to be addressed to design an overall process. Since the organic solvent is also typically the same solvent used in the bulk active manufacturing process, it may already be an approved chemical for the facility. These criteria of simplicity and availability are definitely driving forces for the use of solvents in bulk active facilities.

Issues with solvents include the cost of the solvent. Solvents are usually used "as is," so the cost per kilogram as purchased is the cost of the cleaner. An additional cost includes the cost of reclaiming or disposing of the solvent. It should be remembered that one of the key incidents that brought the Food and Drug Administration (FDA) to its present stance on cleaning validation involved the improper reclaiming of solvents used for cleaning. Care must be exercised so that adequate specifications are in place to qualify the use of reclaimed solvent. Safety in the use of solvents, either in terms of the flammability of the solvent vapors or in terms of occupational exposure to solvent vapors, may also be an issue. Finally, most large pharmaceutical companies have a corporate objective of reducing solvent use and emissions. This is certainly a driving force to reduce the use of solvents in cleaning operations where it is practical.

Aqueous Cleaning

Aqueous cleaning consists of cleaning with water with or without a variety of functional components [11].

Below is a discussion of the types of functional components that might be present in an aqueous cleaning system.

Water

It would not be called aqueous cleaning without water. Water serves as a solvent and as a medium for other functional processes, including hydrolysis, emulsification, and dispersion. Water typically makes up more than 95 percent of the actual cleaning solution.

Surfactants

"Surfactant" is short for "surface active agent." Surfactants used for cleaning generally have a hydrophilic ("water loving") polar end and a lipophilic ("oil loving") nonpolar end. The function of a surfactant is for wetting surfaces (of both the residue and the surface to be cleaned), solubilization, emulsification, and dispersion. Different surfactants may be more effective in providing one of these functional roles. As broad categories, surfactants may be divided into nonionics (those surfactants with no charge), anionics (those surfactants with a negative charge on the polar end), cationics (those surfactants with a positive charge on the polar end), and amphoterics (those surfactants with either a positive or negative charge, depending on the surrounding pH, on the polar end). Nonionics are typically the best choice for emulsification, while anionics are better for wetting and dispersion. Regardless of the surfactants, other features such as foaming need to be balanced with performance issues. In general, low foaming products are preferred for automated cleaning [12].

Chelants

Chelants are products like EDTA (ethylenediamine-tetraacetic acid), NTA (nitrilo triacetic acid), and certain polyphosphates (like sodium hexametaphosphate) that chelate or tie up certain metal ions in aqueous solution. Chelants can be important for any cleaning operation where hard water ions (calcium and magnesium) are present. These ions may be present because unsoftened water is used for cleaning or in the residue to be removed. Such ions are known to interfere with the cleaning process, making the detergent system less effective. The presence of chelants may also help remove trace amounts of iron from the system, thus reducing any tendency for a stainless steel system to rouge.

Solvents

In this context, solvents refer to certain water-miscible solvents, such as glycol ethers. Glycol ethers typically assist in the solubilization of oily or greasy residues.

Bases

Bases include hydroxides such as sodium hydroxide or potassium hydroxide. Bases are used to raise the pH, thus rendering certain acid residues more soluble. They also can assist in the hydrolysis of esters or amides. Finally, bases assist in potentiating surfactants so that the detergency is improved. As a general rule, potassium hydroxide is preferred to sodium hydroxide because it is more "free rinsing"; this may be related to the fact that potassium salts are more soluble than sodium salts.

Acids

Acids include weak to medium strength acids such as glycolic acid (also called hydroxyacetic acid), phosphoric acid, and citric acid. Strong mineral acids such as hydrochloric acid and sulfuric acid are generally not used because of safety reasons or deleterious effects on equipment. Acids are used to lower the pH, thus rendering certain alkaline residues more soluble. They also can assist in the hydrolysis of esters or amides.

Builders

Builders include a variety of alkaline salts, such as trisodium phosphate, sodium silicate, and sodium carbonate. These builders serve to improve the detergency of surfactants. Unless required for specific purposes, builders of these types should be avoided because they may leave insoluble residues. Phosphates and carbonates may precipitate as calcium salts; silicates left behind may dry on surfaces to form residues that are removed only with extreme difficulty.

Dispersants

Dispersants are generally charged, relatively low molecular weight polymers (such as polyacrylates) that assist in suspending solids in water. They are generally used with surfactants, which assist in wetting of solid particles so that they can be effectively dispersed and carried away.

Oxidants

Oxidizing substances, such as sodium hypochlorite, peracetic acid, and hydrogen peroxide, may be used for cleaning to convert insoluble organic residues to shorter chain, more water-soluble fragments. One item to consider in the use of oxidants is that the oxidants selected to improve cleaning may also oxidize surfactants or other organic constituents in the cleaning solution. If this occurs, both the oxidizing advantage of the oxidant and the wetting advantage of the surfactant may be lost. For this reason, care must be used in adding sodium hypochlorite to formulated cleaning products. Formulated chlorinated cleaners commercially sold are usually formulated with surfactants that are stable to hypochlorite bleach. Although oxidants may not be labeled as antimicrobial agents, their oxidizing capability usually provides some measure of antimicrobial activity.

Advantages and Disadvantages of Aqueous Cleaning

The main advantages of aqueous cleaning are cost and environmental acceptability. The costs of cleaning agents for aqueous cleaning (generally used at 1–5 percent diluted in water) are generally less than the cost of pure solvent. While this cost saving may seem great when considered by itself, it should be recognized that switching from solvent cleaning to aqueous cleaning may involve considerable engineering changes. This is particularly so if solvent cleaning involves refluxing the solvent. Aqueous cleaning also has a more "environmentally friendly" image than solvent cleaning. However, there are still issues with aqueous cleaners, particularly relating to the pH of discharges. As a general rule, the discharge of an aqueous stream has to be in the pH 6–9 range. This neutralization is preferably done in a separate holding tank, not in the cleaned equipment, because of the possibility of redepositing residues onto cleaned equipment surfaces due to a pH adjustment.

With aqueous cleaning, there is also more concern about deleterious effects of the cleaning process on the cleaned equipment. For example, with glass-lined vessels, there is concern about the use of highly alkaline cleaners at high concentrations and at high temperatures. Repeated exposure to such conditions may result in etching of the glass surfaces. For stainless steel equipment, the main concern is exposure to high chloride levels, for example, due to hypochlorite-containing cleaners. Repeated use of these under extreme conditions may result in depassivation of the stainless steel and subsequent rouging of the stainless steel surface. In evaluating such effects, it is necessary to consider the actual conditions of exposure of the cleaning agent. Corrosion rates based on lab studies with corrosion rates of 6 mils (0.006 in.) per year based on continuous (24 hours per day, 52 weeks per year) exposure do not translate to a 6-mil loss in one year of actual use; in actual use, the surface may be exposed to the cleaning agent for only a small fraction of time each day or each week.

Safety concerns with aqueous cleaners generally involve accidental contact with either the concentrated or diluted cleaning agent. Such concentrated cleaning agents may be corrosive to the skin or eyes. Dilute (< 5 percent) solutions of such cleaning agents may be eye or skin irritants.

Commodity chemicals such as sodium hydroxide or phosphoric acid diluted in water can be successfully used in cleaning. As compared to formulated specialty aqueous cleaners, commodity chemicals offer significant savings in the cost of the cleaning agent. Such commodities offer an advantage if the primary cleaning mode is based on a pH change with the resultant solubilization or hydrolysis of a residue. However, commodities may result in more cleaning process costs due to longer times required for cleaning. Some companies have also reported shorter rinsing times, for example, using a formulated alkaline aqueous cleaner as opposed to just sodium hydroxide diluted in water. This may be due to the effect of the surfactants (wetting agents) in assisting the rinsing process.

While the cost as purchased per kilogram may be higher for the formulated cleaner, the overall process cost may be less due to time savings. Formulated products may also be more flexible in use because they are multifunctional—the variety of functional components provides cleaning effectiveness over a broader range of residue types that might be present in cleaning in finished drug manufacture or biotech processing. This is based on the fact that in situations where there are many chemical types of residues (for example, both the drug active and a variety of excipients in finished drug manufacture), a multifunctional cleaning solution will be more effective than a cleaning solution that depends on, for example, sodium hydroxide alone. The same analogy applies to the use of a formulated aqueous cleaner in a multiproduct facility. A multifunctional cleaning agent offers a higher probability of having one cleaning solution effective on the variety of product (soil) types that are made on one given piece of equipment. Such an approach may simplify the validation process in a multiproduct facility because one cleaning process used for all products may more easily allow for grouping strategies for cleaning validation work.

One of the concerns with formulated cleaning agents is the selection process. There are a variety of different formulated alkaline cleaning agents on the market. They are certainly not all interchangeable because they have different surfactants and different functional additives. Selection usually has to be done in a laboratory-based, trial-and-error Edisonian approach; such studies will be described in more detail in Chapter 6.

Combination Cleaning Processes

A combination cleaning process involves cleaning with what otherwise might be considered two (or more) separate cleaning processes. One common example of a combination process involves first cleaning with an acidic detergent solution and rinsing, which is then followed by cleaning with an alkaline cleaning detergent and rinsing. In such a strategy, the cleaning process is only complete after the second rinse is complete. Such a strategy might be used if an alkaline detergent is effective in removing one part (e.g., the excipients) of a finished drug, and an acid detergent is more effective in removing other portions (e.g., the active) of that finished drug. Another case where it might be used involves multiproduct equipment, in which some of the products manufactured in the equipment are cleaned with an acid detergent and other products made in the same equipment are more effectively cleaned by an alkaline cleaner. For possible grouping (see Chapter 7), all products are cleaned by the same combination cleaning process.

Care should be exercised in adopting a combination strategy. If a given soil (i.e., manufactured product) is effectively cleaned with an alkaline cleaner alone, it may not be effectively cleaned by the combination process involving exposure first to the acid detergent with subsequent exposure to the alkaline detergent. The reason is that exposure of the soil to the acid detergent may change the nature of the soil so that it is no longer effectively cleaned by the alkaline detergent. This can be evaluated in the laboratory to confirm there are no interactions that would invalidate the use of a specific combination cleaning process.

A second type of combination cleaning process is more common in bulk manufacture. This involves first cleaning with an aqueous-based detergent to, at a minimum, remove the major part of the soil on equipment surfaces. Following rinsing with water, the system is then cleaned with a solvent. The solvent is used for one or more of three reasons: (1) a "polishing" step to further reduce residues on surfaces, (2) a drying process to remove water from the system (particularly if water would interfere with any subsequent processes in the equipment), and (3) a sampling tool to capture residues in the process equipment that are then measured by a suitable analytical technique.

While combination procedures may seem more time consuming and more complex, they may be overall more effective than a simple cleaning process.

REFERENCES

1. Rosen, M. J. 1978. Detergency and its modification by surfactants. In *Surfactants and interfacial phenomena.* New York: John Wiley and Sons, Inc., pp. 272–293.

2. Morgan, J. J., and W. Stumm. 1998. Water. In *Encyclopedia of chemical technology,* 4th ed., vol. 25. New York: John Wiley and Sons, pp. 383–388.

3. Rosen, M. J. 1978. Solubilization by solutions of surfactants: Micellar catalysis. In *Surfactants and interfacial phenomena.* New York: John Wiley and Sons, Inc., pp. 123–148.

4. Rosen, M. J. 1978. Emulsification by surfactants. In *Surfactants and interfacial phenomena*. New York: John Wiley and Sons, Inc. pp. 224–250.

5. Rosen, M. J. 1978. Dispersion and aggregation of solids in liquid media by surfactants. In *Surfactants and interfacial phenomena*. New York: John Wiley and Sons, Inc., pp. 251–271.

6. Rosen, M. J. 1978. Reduction of surface and interfacial tension by surfactants. In *Surfactants and interfacial phenomena*. New York: John Wiley and Sons, Inc., pp. 149–173.

7. Rosen, M. J. 1978. Wetting and its modification by surfactants. In *Surfactants and interfacial phenomena*. New York: John Wiley and Sons, Inc., pp. 174–199.

8. Tau, K. D., V. Elango, and J. A. McDonough. 1994. Ester, Organic. In *Encyclopedia of chemical technology*, 4th ed., vol. 9. New York: John Wiley and Sons, pp.783–786. (1994).

9. Lynn, J. L. Jr. 1993. Detergency. In *Encyclopedia of chemical technology*, 4th ed., vol. 7. New York: John Wiley and Sons, pp.1081–1087.

10. LeBlanc, D. A., D. D. Danforth, and J. M. Smith. 1993. Cleaning technology for pharmaceutical manufacturing. *Pharmaceutical Technology* 17 (7): 84–92.

11. Lynn, J. L. Jr. Detergency. In *Encyclopedia of chemical technology*, 4th ed., vol. 7. New York: John Wiley and Sons, pp. 1073–1081.

12. Lynn, J. L. Jr. and B. H. Bory. 1997. Surfactant. In *Encyclopedia of chemical technology*, 4th ed., vol. 23. New York: John Wiley and Sons, pp. 478–541.   .

3

Cleaning Methods

The chemical cleaning agents discussed in Chapter 2 are very important for cleaning performance. However, equally important is the cleaning method used. Cleaning methods are usually differentiated based on the extent of disassembly required for the cleaned equipment and on the method of contacting the chemical cleaning agent with the surface to be cleaned. This chapter will cover different cleaning methods as well as some key engineering factors in the design of equipment to facilitate ease of cleaning. Knowing of the limitations of cleaning methods is important to facilitate appropriate controls to help assure a validated or validatable cleaning process. Cleaning application methods discussed in this chapter include the following:

- CIP (clean-in-place)

- Agitated immersion

- Static immersion (soaking)

- Automated parts washing

- Ultrasonic cleaning

- High pressure spraying

- Manual cleaning

The features, advantages, and disadvantages of each of these methods will be covered separately.

CLEAN–IN–PLACE

Cleaning applications can be classified into two groups based on the extent of disassembly—CIP (clean-in-place) and COP (clean-out-of-place). Technically, CIP applies to all situations in which disassembly is not performed and can apply to situations in which process equipment is merely flooded with cleaning solution. As a practical matter, CIP has usually come to mean an automated cleaning system involving spray devices to distribute the cleaning solution to all process vessel surfaces [1,2,3,4].

In the ideal world, CIP involves no disassembly of the equipment prior to initiation of the cleaning cycle. This means that the equipment is designed with CIP in mind—it is "prepiped" to handle the introduction of chemical cleaning solutions into the equipment, the flow of that cleaning solution through the equipment to contact and clean all surfaces, the discharge of the spent cleaning solution to drain, the rinsing of all cleaned surfaces, and, optionally, the drying of the cleaned surfaces. CIP systems are usually designed for cleaning with aqueous cleaning solutions. The key parts of the CIP system are the spray device(s), the

CIP unit, and the associated piping to carry the solution to and from the equipment to be cleaned. A typical CIP system schematic is illustrated in Figure 3.1, and an actual CIP unit is shown in Figure 3.2.

The CIP unit consists of one or more CIP tanks, a recirculation pump, and a process control unit. The concentrated cleaning agent storage tank is usually a plastic tank or carboy, although a drum of the cleaning agent may serve as the vessel for the concentrated cleaning agent. A separate CIP wash tank (also called the recirculation tank) is used for cleaning agent dilution and serves as the reservoir of diluted cleaning agent for recirculation throughout the CIP system. The recirculation pump is usually a centrifugal pump that pumps the cleaning solution from the wash tank to the spray devices. Usually the spray device (a device not unlike a shower head, albeit one that typically sprays up rather than down) sprays the cleaning solution around the dome of a process vessel so that the cleaning solution is adequately distributed within the process vessel. Different spray devices are illustrated in Figure 3.3. Spray devices are usually of either fixed (sometimes called stationary) or rotating (sometimes called dynamic) types. Fixed spray devices (usually spray balls) are mounted in one orientation, and the direction of spray does not change during the cleaning process. Fixed spray devices are simple in design, have no moving parts, are free draining, and are therefore ideal for "sanitary" applications. Fixed spray devices are generally used at lower pressures of about 15–40 psi. Fixed spray devices may be permanently installed in a tank, or they may be installed in place immediately before CIP cleaning is performed.

Rotating spray devices are also mounted in one orientation; however, the head will rotate through a

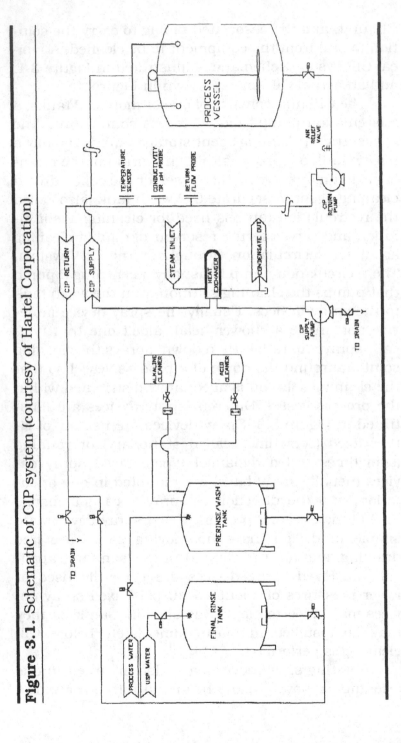

Figure 3.1. Schematic of CIP system (courtesy of Hartel Corporation).

Figure 3.2. Typical fixed CIP unit (courtesy of Martin Petersen Company, Inc.).

pattern providing direct impingement over a larger area. Rotating spray devices are generally operated at higher pressures, up to 100 psi or greater. The higher pressure allows more mechanical energy to be inputted into the system, therefore making the cleaning process more efficient in some cases. Because of the rotation and the complexity, these spray devices are usually not free draining and are not sanitary

Figure 3.3. Representative fixed CIP spray devices
(courtesy of Hartel Corporation).

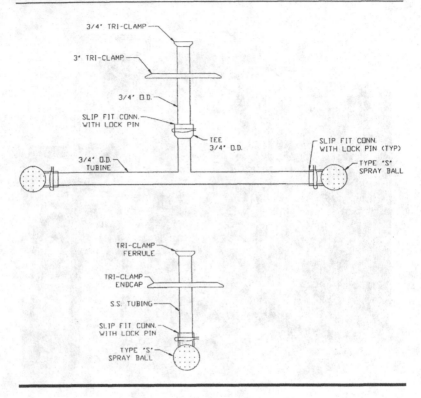

(although there are some designs described as sanitary). Therefore, they are used in systems in which
the spray head is removed from the system after
cleaning. They must be reinstalled for the next cleaning procedure. Depending on the number and location of "inserts" within a process vessel, one or more
spray devices may be needed. The location and spray
orientation of spray devices may be near the dome.
Fixed spray devices will spray up onto the dome (to
get good overall coverage) but may also have to be

strategically placed to assure adequate cleaning and rinsing.

After the cleaning solution is distributed within the vessel, the cleaning solution cascades down the sides of the process vessel, across the vessel bottom, and down the discharge pipe. For once-through CIP systems, the cleaning solution is not reused; the discharge pipe leads to either a holding tank (for chemical neutralization) or directly to a waste treatment facility. For recirculating systems, the used (but not saturated) cleaning solution returns to the CIP unit and is then pumped back through the spray devices to form a recirculating loop. It should be noted that for prewashes (sometimes called prerinses) and for any rinsing done after circulation of the cleaning solution, the CIP system is operated in the once-through mode. In other words, the terms *once-through* and *recirculation* actually describe only the circulation of the cleaning solution. For prewashes and for rinses (both for a recirculation system and a once-through system), water is usually passed through the spray devices and exits to the discharge pipe; it does not recirculate through the CIP unit.

Once-through systems are rarely used for CIP cleaning, primarily because of the cost of the cleaning solution required (both the cost of the water and the cost of the cleaning chemical). While recirculating systems are most common, the disadvantage is that the residues being cleaned may come in contact with the CIP tank, the associated CIP piping, and the spray devices. This means that residues may be left behind on these surfaces. When one evaluates the potential for cross-contamination, these surfaces in the CIP unit should be considered as potential sources of contaminating residues. To prevent redeposition of residues in the CIP cleaning solution tanks, it may be necessary to include a spray device in the CIP tank. This spray device would effectively clean and rinse all surfaces of the storage tank.

Fixed CIP systems are those that are "hard piped" into the manufacturing equipment. There are also portable CIP systems, which are typically skid mounted and may be moved from place to place within a facility. These portable CIP systems are less expensive and can be used for many process vessels. The disadvantages of portable systems are that they always require some assembly and disassembly (attaching hoses). With either fixed or portable systems, there may be other disassembly required prior to performing CIP. For example, a filter housing may be opened, the filter may be removed, the system closed again, and then a CIP cycle performed. Also, the process vessel to be cleaned may be opened to install spray devices in the vessel.

The advantages of CIP systems are as follows:

- *Designed for cleanability:* Ideally, the CIP system is designed into the equipment to be cleaned. This not only involves good design for the CIP system but also good design for the equipment to be cleaned. Some of these design considerations will be covered in Chapters 4 and 5. Retrofitting existing process vessels to be cleaned with a CIP system can be effective, but it may result in either equipment modifications or over engineering of the CIP system to account for proper cleaning.

- *Automated:* CIP systems can automate some or all parts of the cleaning process. This includes automating parameters such as concentration of the cleaning solution, temperature, and times of the various cycles (prerinse, wash, rinse, and dry). CIP systems can also record various process parameters, such as temperature, cycle times, and pressures, that can help to monitor process performance.

- *Consistency:* Because there is a minimum of operator intervention in a CIP process, there is less chance for variability in process parameters, unlike manual cleaning in which there is significant human intervention. This has to be balanced with the observation that a bad CIP cycle (e.g., with poorly chosen process parameters) will *consistently* give bad results. While CIP minimizes human intervention at the operating stage, it requires significant human intervention at the design phase.

- *Water/cleaner savings:* CIP systems can minimize (as compared to agitated immersion methods) the use of the cleaning solution because the process vessel is not flooded with the cleaning solution. Rather, only enough cleaning solution is used so that the cleaning solution completely wets and covers all surfaces of the vessel (as well as completely wetting and filling all associated piping on the supply and discharge sides). As compared to completely filling the vessel with cleaning solution (as in agitated immersion), much less water and chemical cleaning agent are used. While this savings may be significant for a recirculating system, water and cleaning agent savings for once-through CIP processes are much less; the usage in a once-through process may be even more than in an agitated immersion process.

- *Time savings:* The time savings possible here include lower total washing and rinsing cycle times, due to lower filling and dumping times as compared to an agitated immersion process, and less time for disassembly and reassembly of the equipment as compared to COP processes.

- *Equipment wear:* The lack of disassembly/reassembly may also reduce wear and tear on the equipment. This may be significant for glass-lined vessels and delicate parts.

- *Ease of validation:* Two factors contribute to the ease of validation: automation, which assures more consistent performance, and the lack of assembly/disassembly, which minimizes disruptions.

- *Safety of operators:* Because CIP systems involve little operator intervention, they may be perceived as safer systems. This has to be balanced against the fact that more aggressive cleaning agents are used in CIP systems; therefore, when accidental contact with the cleaning agent occurs, the effects may be more serious.

While these advantages are significant, there are also disadvantages to CIP systems.

- *Lack of flexibility:* There is flexibility within the parameters designed into the system, but it may be difficult to make major adjustments once a system is installed.

- *High initial capital cost:* CIP systems cost more in capital than other cleaning systems.

This may be viewed as minor if included as part of the initial equipment purchase. However, when viewed alone in the retrofitting of existing equipment, the cost may be viewed as excessive (as compared to other alternatives).

- *Use of more aggressive cleaning agents:* Because most CIP systems in the pharmaceutical industry are fixed spray ball, low-pressure systems, the major cleaning effect is due to the chemical cleaning agents. Physical or mechanical effects due to impingement of the cleaning solution or flow of the cleaning solution are less important. Therefore, more aggressive (more alkaline, more acidic) cleaning agents may be used. These may be of more concern both in terms of deleterious effects on equipment during normal use and safety issues involving workers with accidental exposure.

Overall, the advantages of CIP systems usually outweigh any disadvantages provided the system can be initially engineered for CIP. For existing manufacturing equipment, the main issue is whether the equipment can be readily modified and retrofitted for CIP.

AGITATED IMMERSION

Agitated immersion involves filling a process vessel (and associated piping) with a cleaning solution and then agitating the solution with a mechanical agitator already in place in the process vessel. The cleaning solution can be either an organic solvent or an aqueous cleaning agent. In such a system, after the cleaning step, the vessel is drained and then filled multiple times with rinse water (or other rinsing solvent) to remove soil residues and the cleaning solution. During the rinsing steps, the rinse solutions are agitated just as in the cleaning step. It should be noted that pumping a cleaning solution, for example, from the discharge pipe of a vessel, through a recirculation pump, and back to the top of the vessel, generally will not be very effective in achieving agitation. In such a system, it is difficult to achieve adequate flow within the larger vessel. There are likely to be significant areas with low flow rates. In such an example, the performance would be closer to the performance expected for static immersion (discussed below).

In agitated immersion cleaning, the major cleaning effect is due to the chemical action of the cleaning agent. The effect of the agitation is to bring fresh cleaning solution in contact with the residue on the surface. This maximizes the cleaning effect. The second purpose of the agitation is to remove solubilized/emulsified/suspended soils from the equipment surfaces and carry them to the bulk cleaning solution. This prevents soil concentration gradients from forming near the equipment surfaces; such gradients can slow down the cleaning process.

Agitated immersion can be performed under a variety of process conditions, including various times, temperatures, cleaning agents, and flow conditions. The emphasis is contact of the agitated cleaning solution with all surfaces. The advantages of agitated immersion systems are as follows:

- *Low capital cost:* Very little is needed to retrofit systems for agitated immersion. All that is needed is a system for feeding the cleaning agent into the water used for cleaning.

- *Simplicity:* For equipment that can be flooded with cleaning agent solution, agitated immersion is a simple system. There are few variables to evaluate. Once the cleaning solution and temperature are chosen, the only variable is time (variations in the rate of agitation are generally not significant).

The disadvantages of agitated immersion are as follows:

- *Process time:* If one is cleaning a 5,000 L vessel in a CIP application, one can begin pumping cleaning solution into the system and have all surfaces wetted in a matter of a few minutes. With agitated immersion, it is necessary to fill the vessel with 5,000 L of cleaning solution before the "time clock" on cleaning can begin. In addition, the vessel has to be drained of the 5,000 L at the end of the cleaning step. The added times are also applicable to the rinsing step, of which generally there are more than two. These times will significantly add to the cleaning cycle time. Note that the time for the cleaning step itself will be approximately the same as in a CIP cycle; what are different are the filling and draining times.

- *Water and cleaning agent use:* Compared to a recirculating CIP system, an agitated immersion system will use approximately 200–800 percent more water and chemical cleaning agent. The operating costs in terms of time and materials will therefore be higher than with a recirculating CIP system.

- *Equipment limitations:* If the process equipment does not have a built-in agitator, then it may be difficult to perform agitated immersion cleaning. In addition, equipment design may limit or make agitated immersion difficult. For example, it may be very difficult to completely flood equipment with a dome. Removing the last amount of air from the dome may require a special bleed valve. Unfortunately, with agitated systems, inadvertent splashing may cause problems with residues on the dome, so removal of all air is necessary for adequate cleaning.

Agitated immersion may be an acceptable alternative if an adequate CIP system cannot be installed because of design or cost considerations.

STATIC IMMERSION

Static immersion is agitated immersion without the agitation. In static immersion, a process vessel is merely flooded with the cleaning solution (either aqueous or organic solvent). The sole cleaning effect is due to the chemical action of the cleaning solution. Even this chemical cleaning action is minimized because concentration gradients of residues solubilized, emulsified, or suspended near the equipment surfaces tend to retard the cleaning process. Depending on the nature of the residue and the cleaning agent, times for the cleaning step can take from 50 percent longer to as much as 500 percent longer than with agitated immersion. Static immersion is generally a cleaning method of last resort.

The advantages of static immersion systems are as follows:

- *Low capital cost:* Very little is needed to retrofit systems for static immersion. All that is needed is a system for feeding the cleaning agent into the water used for cleaning.

- *Simplicity:* For equipment that can be flooded with cleaning agent solution, static immersion is a simple system. There are few variables to evaluate. Once the cleaning solution and temperature are chosen, the only variable is time.

The disadvantages of static immersion are as follows:

- *Process time:* If one is cleaning a 5,000 L vessel in a CIP application, one can begin pumping cleaning solution into the system and have all surfaces wetted in a matter of a few minutes. With static immersion, it is necessary to fill the vessel with 5,000 L of cleaning solution before the "time clock" on cleaning can begin. In addition, the vessel has to be drained of the 5,000 L at the end of the cleaning step. The added times are also applicable to the rinsing step, of which generally there are more than two. These times will significantly add to the cleaning cycle time. In addition, the time for the cleaning step and for the rinsing steps will be significantly longer than with agitated immersion due both to the lack of replenishment of fresh cleaning solution at the equipment surfaces and to lower forces acting on that surface.

- *Water and cleaning agent use:* Compared to an agitated immersion system, a static immersion system will use approximately the same amount of cleaning agent but probably more water due to the fact that additional rinses may be necessary to insure adequate removal of cleaning solution residues from all surfaces. The operating costs in terms of time and materials will therefore be higher than with an agitated immersion system.

- *Equipment limitations:* The same issue of equipment design for agitated immersion applies here. For example, it may be very difficult to completely flood equipment with a dome. Removing the last amount of air from the dome may require a special bleed valve. This may or may not be a problem, depending on whether the dome is a critical surface requiring cleaning.

Static immersion may be an acceptable alternative only as a matter of last resort. Other alternatives should probably be exhausted before this alternative is chosen.

AUTOMATED PARTS WASHING

Automated parts washing is in the domain of true COP applications [5]. Parts that are disassembled from process equipment or other small parts such as scoops that are used in the manufacturing process can be cleaned in this manner. The parts are placed in a mechanical washer and processed through cleaning, rinsing, and drying cycles. The cleaning solution and the rinse water are applied to all surfaces of the objects to be cleaned by spray jets and nozzles. Automated washers are of two types. In *cabinet washers* (see Figure 3.4), the parts are placed on racks inside the washing cabinet, and all cycles occur with the racks in a fixed position (similar to a home dishwasher). In *tunnel washers* (see Figure 3.5), the parts are placed on racks that travel through the tunnel washer. Cleaning, rinsing, and drying occur at different locations within the tunnel washer (similar to some commercial car wash facilities). The degree of controls on a washer can vary considerably. With some washers, there are minimal controls and recording of data. With so-called "GMP" (Good Manufacturing Practice) washers, there is considerably more documentation to enable the washer to be more easily validated.

The performance of the automated washer is due to both the mechanical impingement of the spray systems and to the chemical action of the cleaner used. With the systems discussed previously (CIP, agitated immersion, and static immersion), the chemical cleaning agent has been significantly more important than any mechanical effects. In automated parts washers, the mechanical cleaning effects are at least as important as the cleaning agents used. In parts washers, spray pressures may be only in the 5–20 psi range. The spray is designed so that all of the parts (but not necessarily all surfaces of a given part) obtain adequate spray impingement. For this reason, issues like clogged spray nozzles can significantly affect cleaning performance. Because of the spray pressure, the issue of detergent foaming is significant. Foam is very difficult to rinse from cleaned parts, thereby increasing the possibility of leaving detergent

Figure 3.4. Typical cabinet (rack) washer (courtesy of STERIS Corporation).

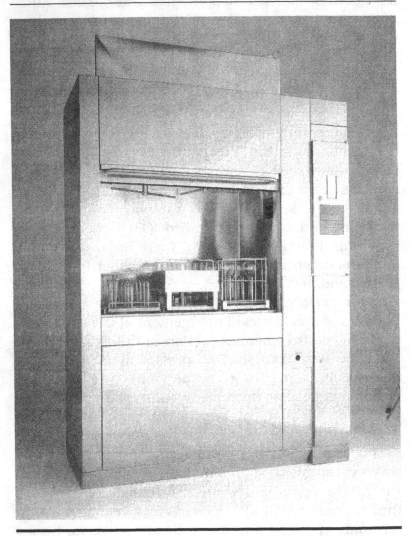

residues behind. Foam can also "hang up" in various parts of the washer and redeposit on cleaned parts after the rinsing or drying step. For this reason, it is important that the rinsing be designed to rinse both the cleaned parts *and* the internal equipment surfaces.

Figure 3.5. Typical tunnel (continuous) washer (courtesy of STERIS Corporation).

Careful selection of the cleaning agent is required for automated parts washers.

Automated parts washers have a number of advantages for smaller items:

- *Consistent performance:* Because the process is automated, the cleaning process is more consistent and thereby more easily validatable. As with any automated system, initial selection of a cleaning cycle (including detergent system and process parameters) is necessary to insure *consistently good* performance.

- *Time savings:* Once the items are loaded into the automated washer, all of the parts are washed simultaneously. For comparison, in a manual sink washing situation, each part has to be washed sequentially. In addition, the cleaning can be done at higher temperatures, which also shortens the cleaning time. The entire process can be accomplished in a parts washer for the same or less elapsed time with fewer man-hours.

- *Chemical and water savings:* The operating costs (chemical cleaning agent, water usage) of an automated washer system is significantly less than with a manual sink cleaning operation.

- *Safety:* With an automated parts washer, the exposure of workers to hot, cleaning agent solutions is minimized.

The disadvantages of automated parts washers are as follows:

- *Initial capital cost:* The cost of a good quality parts washer is significantly higher than equipment for other manual cleaning procedures.

- *Unsuitable for delicate parts:* Because of the mechanical energy imparted in the cleaning process, delicate parts may be damaged in the cleaning process.

Automated parts washing is the preferred method for washing small parts (except for delicate parts) because of labor savings and the consistency of performance.

ULTRASONIC WASHERS

Ultrasonic washers are essentially open or covered tanks with a cleaning solution in the tank. Tanks may be covered to prevent evaporation or aerosolization of the cleaning solution. The tank is equipped with ultrasonic transducers, which pass high frequency sound waves through the cleaning solution [6]. At the surfaces of objects placed in the ultrasonic cleaning solutions, the sound waves cause tiny bubbles to form. As these bubbles grow, they eventually collapse upon themselves. The mechanical energy of the collapse helps dislodge any residues or particulates on the surface. This process is called *cavitation.* The cleaning agent solution helps to wet the residue and then keep it suspended or emulsified. As a general rule, ultrasonics are run at temperatures below about 55°C. Ultrasonic baths come in different sizes, from 1 L up to 20 L or more. The ultrasonic process is most appropriate for cleaning delicate parts that might be damaged by other cleaning processes or for parts with small orifices or other openings. For example, filling needles might best be cleaned by an ultrasonic process. Following the exposure in the ultrasonic bath, the part is manually removed and rinsed by any suitable method, such as with flowing water. For small orifices, rinse water may have to be pumped or aspirated through the orifice.

The advantages of ultrasonics are as follows:

- *Excellent cleaning for delicate items:* Ultrasonics are the cleaning process of choice for these items.

- *Low initial capital cost:* Compared to automated parts washers, simple ultrasonics are relatively inexpensive. It should be noted that automated transport systems for parts may add significantly to the cost.

The disadvantages of ultrasonics are as follows:

- *Significant manual processing:* Unless automatic transport and unloading/loading systems are used, an ultrasonic system requires significant handling of the part. This also includes a requirement for a separate rinse process. The process cost, particularly labor, is typically more than with an automated parts washer.

- *Validation issues:* Because the transducers lose power over time (as they age), the validation should be done under conditions that will account for this loss of power over time. As a practical matter, the system is continually varying.

Overall, its performance for small, delicate parts makes it the method of choice for these items.

HIGH–PRESSURE SPRAYING

A high-pressure spray application involves using a high-pressure, continuous, directed water or detergent solution to clean parts or to clean the inside of process equipment. Water pressures may be on the order of hundreds of psi. This usually involves the use of a spray "wand" of some type. In its simplest version, the spray wand is manually controlled to clean small parts. In a more sophisticated version, an automated spray wand is inserted inside process vessels, where the spray direction and orientation are robotically controlled to clean and rinse all surfaces.

High-pressure spray applications involve a more significant contribution from mechanical energy inputted from impingement of the cleaning solution on the surface. This serves to dislodge any residues on the surface. While water alone would be effective in dislodging residues, a detergent solution is usually needed to wet the residue and to assist in suspending and/or emulsifying it so it is not redeposited in other parts of the system. The selection of detergent will depend on the residues being cleaned. The time of cleaning is determined by the operator or by the preprogrammed system for automated systems. The temperature of the water is usually hot (> 50°C). The same high-pressure system, with water alone, is used to rinse the parts or the vessel interior. Because of the possibility of "splash back" from the high-pressure detergent solution, extra care should be used in considering the safety of the operator, particularly eye safety.

The advantages of high-pressure spray applications are as follows:

- *Relatively low capital cost:* This is true for most manual systems. For the automated, robotic systems for vessel interiors, the cost may be significant.

- *Highly effective:* At higher pressures, the impingement provides significant energy input to clean most surfaces.

The disadvantages of the high-pressure systems are as follows:

- *Large water use:* Because these are generally once-through applications (the solution is sprayed onto the object, and then the solution goes directly to the drain), a large amount of water is used.

- *Equipment limitations:* This method is good for the inside of vessels (provided shadow areas are accounted for) or for smaller parts that can be sprayed in a dedicated spraying area. It is not suitable for delicate parts.

- *Variability of manual systems:* With automated systems, the spray direction, orientation, and time are controlled. In manual systems, it is more difficult to obtain consistent results.

High-pressure applications are effective where all parts of equipment can be directly sprayed.

MANUAL CLEANING

Manual cleaning covers a variety of cleaning types. The types of manual cleaning covered are wiping, sink brushing, and equipment brushing. They all have the inherent advantage of low capital costs and the inherent disadvantage of higher variability. When done right, they can be very effective. Manual methods of cleaning also allow operators some degree of immediate feedback on their cleaning performance.

Wiping

The simplest manual cleaning is *wiping* with a lint-free cloth of some kind and using a cleaning solution, for example, the manual wipe-down of a tableting press. Wiping depends to a large extent on the mechanical energy input by the person doing the wiping. The effect of the cleaning agent is mainly to wet soils and to keep them from redepositing on the surface. Since most of the cleaning is coming from mechanical energy, and because of the safety concerns of using highly aggressive cleaning agents in a manual operation, the cleaning agents used are generally neutral, mildly alkaline, or mildly acidic. Depending on the application, an alcohol solution in water may be used. The cloth used is a knit or nonwoven, low-linting fabric. A key issue in manual wiping processes is to insure 100 percent coverage of the surface to be cleaned. For this reason, wiping should be done with overlapping strokes. If the surface is visibly soiled, then the operators can easily see where they have wiped. A second issue in wiping is the need for rinsing. Certainly all product contact surfaces cleaned with detergent solutions should be rinsed. Rinsing may not be needed if cleaning is done with an alcohol/water solution alone (in that case, the alcohol/water wipe may be repeated to assure the removal of residues).

The advantages of manual wiping are as follows:

- *Simplicity:* Wiping can be used on any surface that one can physically reach with one's hands. There is immediate feedback on relative performance (as least from a visual standpoint).

- *Flexibility:* A variety of wiper and cleaning agent types can be used. With wiping, it is possible to focus the cleaning operation on a specific item without having to clean any adjacent items.

- *Low cleaning agent cost:* Since only enough cleaning agent is used to wet the wiper and the surface, the usage is very low.

The disadvantages of manual wiping are as follows:

- *Inherent variability:* Performance depends to a large extent on such items as pressure applied in wiping. Training for operator consistency is difficult.

- *High labor cost:* The unit labor cost will depend on the surface area cleaned. There is no way to automate this process.

Wiping is the method of choice where surfaces are readily accessible, where equipment cannot be disassembled, or where the absence of a floor drain or the necessity to protect the general cleaning area does not make spraying feasible.

Sink Brushing

What more can be said about this method? This is closest to manually washing dishes at home. Parts are taken to a sink, placed in a detergent solution, manually scrubbed using a brush or abrasive pad, and then rinsed under flowing water. This method has the advantage of allowing for an extended soaking time (to wet and loosen the soil) in the warm ($< 50°C$) detergent solution prior to manually brushing it from the surface of the part. The detergent solution is typically a neutral or mildly alkaline detergent, selected because of the greater potential of skin or eye contact in this type of manual cleaning. Cleaning performance depends to a large extent on the consistency (time and pressure) of the brushing action. The keys for validation of this process are control of temperature, consistency in brushing, and control of any presoaking time. A carefully written cleaning SOP, along with adequate training and re-training, are required.

The advantages of sink brushing are as follows:

- *Simplicity:* This method can be used on any part that can fit into the sink. There is immediate feedback on relative performance (as least from a visual standpoint).

- *Flexibility:* A variety of brushes or abrasive pads can be used. Soaking or brushing times can be adjusted depending on the residues on the surfaces and the surface configuration.

The disadvantages of sink brushing are as follows:

- *Inherent variability:* Performance depends to a large extent on such items as soak time, application pressure, and brushing time. Training for consistent performance is difficult.

- *High labor cost:* The unit labor cost will depend on the number of items processed. The only way to automate this is to use an automated parts washer.

This is the method of choice where disassembled or small parts cannot be washed in an automated parts washer and where the parts are rugged enough to withstand this process.

Equipment Brushing

Equipment brushing is similar to sink brushing in that a detergent solution is introduced into the interior of a process vessel. The detergent solution is manually applied to all surfaces with a brush (usually on a long handle). In some cases, the brushing is performed through a manway at the top of the equipment. In other cases, operators may enter the vessel (under appropriate lockout procedures) and clean the equipment from the inside. This latter approach is generally avoided from a safety perspective. Equipment brushing differs from sink brushing in that only enough detergent solution is introduced into the vessel to provide enough solution to cover all surfaces by brushing (that is, the vessel is not flooded with the cleaning solution). Following brushing of all surfaces, the detergent solution is drained, and the vessel is rinsed with water using a hose. Equipment brushing is used when most other alternatives are not feasible or practical.

There is only one advantage of equipment brushing:

- *Flexibility:* The cleaning effort (the brushing) can be focused on those areas of the equipment that require that extra effort.

The disadvantages of equipment brushing are as follows:

- *Equipment limitations:* Brushing from outside the equipment may not be possible because of equipment geometry. Brushing from the inside has significant safety concerns. Brushing from the inside also lengthens the down time because of lockout procedures for individuals to enter process vessels.

- *High labor cost:* The cost is dependent on the preparation time and the surface area to be cleaned.

SELECTING A CLEANING METHOD

No one cleaning method is inherently better than another for validation and cleaning purposes. The advantages and disadvantages of the different methods have to be evaluated and compared. Consistency alone is not adequate for selecting a cleaning method. Consistency *and* performance are required to produce a practical and achievable cleaning method. In addition, the selection of a cleaning method must be integrated with the selection of the cleaning agent(s), both of which must be considered in light of the residue(s) to be removed. An overall systems approach to cleaning can more effectively assist in the selection of the proper cleaning method for any process.

REFERENCES

1. Stewart, J. C. and D. A. Seiberling. 1996. Clean in place. *Chemical Engineering* 103 (1): 72–79.

2. Seiberling, D. A. 1992. Alternatives to conventional process/CIP design for improved cleanability. *Pharmaceutical Engineering* 12 (12): 16–26.

3. Adams, D. G., and D. Agarwal. 1990. CIP system design and installation. *Pharmaceutical Engineering* 10 (6): 9–15.

4. PDA Biotechnology Cleaning Validation Subcommittee. 1996. *Cleaning and cleaning validation: A biotechnology perspective.* Bethesda, Md., USA: Parenteral Drug Association, pp. 65–84.

5. PDA Biotechnology Cleaning Validation Subcommittee. 1996. *Cleaning and cleaning validation: A biotechnology perspective,* Bethesda, Md., USA: Parenteral Drug Association, pp. 85–87.

6. Fuchs, F. J. 1995. The key to ultrasonics—Cavitation and implosion. *Precision Cleaning* 3 (10):13–17.

4

Process Parameters in Cleaning—Part I

Selecting the cleaning agent, as discussed in Chapter 2, is only part of the process to define a validatable cleaning process. Other cleaning process parameters that also affect the quality of cleaning [1] include the following:

- Time

- Action

- Concentration

- Temperature

- Surface type and quality

- Soil levels

- Soil conditions

- Mixing

- Water quality

- Rinsing

- Environmental factors

The first four issues will be covered in this chapter; the rest will be covered in Chapter 5.

TIME

There are three aspects of time as it relates to cleaning validation. The first and most obvious is the length of time the cleaning solution is in contact with the surfaces to be cleaned. As a general rule, the longer the time of cleaning, the better the cleaning efficiency. With other conditions the same, a cleaning time of 60 minutes will generally give as good or better results than a cleaning time of 30 minutes. Turnaround time in changeovers is critical in many situations, so most manufacturers will select the shortest time to do an adequate and consistent job in cleaning. There may be cases when the cleaning time extends beyond that which is necessary for good cleaning performance. One example would be a case in which products to be cleaned are grouped together and cleaned with one cleaning process. In that case, one of the products may be adequately cleaned in a 15-minute cycle, one in a 30-minute cycle, and another in a 60-minute cycle. In order to group these for cleaning validation purposes, all products may be cleaned with the 60-minute cycle. In extending the cleaning cycle, one concern involves deleterious effects of the cleaning agent and/or cleaning cycle on the processing equipment. Proper selection of the cleaning process and process equipment should minimize this concern.

A second aspect of time is the time from the end of manufacturing until cleaning is performed. This can be critical because the nature or condition of the product to be cleaned (sometimes called the "soil," even though most pharmaceutical manufacturers may be reluctant to consider their manufactured products as soils) may change with time. For example, it may be intuitively obvious that if one completes the manufacturing of a batch of drug product on Friday and then leaves the process equipment un- cleaned over the weekend, it may be harder to clean that process vessel the following Monday morning (as compared to immediately cleaning the process equip- ment on Friday afternoon). In other words, a cleaning process validated on "fresh" soil may not necessarily be effective on "aged" soil. Cleaning will not necessar- ily be more difficult, but it is a concern serious enough such that the Food and Drug Administration (FDA) raised this issue in their cleaning validation guidance document [2]. Things that can occur with the passage of time include the following:

- Drying of the soil, due to evaporation of wa- ter or solvent from the soil, may make the soil more difficult to penetrate by the clean- ing solution.

- Microbes may proliferate in the soil, thus significantly increasing the bioburden to be reduced during the cleaning process.

- A temperature change (usually cooling) re- sulting in the hardening of waxy excipients. These hardened excipients generally will be more difficult to remove unless the tempera- ture is increased again.

For validation purposes, it is incumbent on manufacturers to define in their cleaning SOPs (Standard Operating Procedures) the time limit in which the equipment must be cleaned. The most extreme time (provided it is the worst case) should be included as one of the three validation runs for the Process Qualification (PQ) portion of cleaning validation. That longest time can be selected based on practical manufacturing circumstances. If the production department can commit to insuring that cleaning is started within X hours of the end of process, then X hours (or perhaps X + 1 hours, or even 2X hours) can be selected for the time limit by which cleaning should begin.

What happens if, for whatever reason, the time limit is exceeded in a validated cleaning process? For example, if an SOP specifies that cleaning must be performed within 8 hours of the end of manufacturing, what does one do if cleaning is not initiated until 12 hours after the end of manufacturing? The manufacturer has two options, both of which involve additional testing ("verification" testing) to insure the equipment is adequately clean. One option is to clean with the existing SOP and then test for residues in the same manner (i.e., sites and analytical methods), using the same acceptance criteria as in the three original PQ cleaning validation runs. If the results are acceptable, this is adequate verification that the equipment is adequately cleaned, and it may be justification for extending the time limit in the SOP. If the results are unacceptable, then one must proceed to the second option.

The second option is to do some kind of heroic cleaning, such as significantly extending the cleaning time or cleaning the equipment twice with the same cleaning SOP. Following this heroic cleaning, verification is done. This may involve the sampling and analytical techniques used for acceptance criteria in the original cleaning validation PQ work. Heroic cleaning should continue until acceptable cleaning verification results are obtained. This option does not permit any change in the cleaning SOP time limit based on verification testing. Manufacturers will typically choose this option if it is important to have the equipment turned around in as short a time as possible. Manufacturers may choose the first option if they view the possible extension of the time limit as a desirable modification to their SOP.

The third aspect of time is the time from the end of the cleaning process until the time that the process equipment is used for manufacturing the next product. The issue here is that the cleaned process equipment does not stay clean indefinitely. Just as wrapped surgical instruments sterilized in a hospital autoclave are not considered sterile forever, the cleaned process equipment must have a defined shelf life. There are several possibilities for the recontamination of cleaned equipment. One is microbial proliferation. If cleaned equipment is left with areas where rinse water can pool or hold up, then that water (with proper nutrients) may serve as a means for significant microbial proliferation. This can be minimized by insuring that equipment is properly designed, including pipes sloping (1/16 to 1/8 in. per ft) to drain and proper drain placement. A drying step after cleaning can also help. Drying steps may include flushing with an alcohol solution and allowing the alcohol to evaporate or drying with a heated air or nitrogen purge. This example of microbial recontamination is cited by the FDA in their cleaning validation guideline document [2].

Another example of recontamination is recontamination from the air with dust particles. This can occur if equipment is not properly protected (covered or wrapped) after cleaning. The necessity for covering or wrapping will depend on the time and location of storage. If the equipment is to be used again within a few hours, this concern is minimized (but not totally eliminated). For extended storage, such as more than one week, this issue should be addressed, with equipment appropriately protected from recontamination.

A major concern is how the shelf life of cleaned equipment is established and to what extent lab studies are necessary to support this. This is generally done by evaluating the use of the equipment, how and where it is stored, the routes of recontamination, and the nature of the recontaminating residue. In the simplest cases (cleaned and dried equipment, covered with a nonwoven polyolefin wrapping, and stored in a controlled environment area), it may be possible to justify extended storage with an analysis and summary memo by a scientist qualified to make that judgment. In other cases, it may involve qualification protocols in which equipment is stored under worst-case conditions for a defined time period. At the end of that time period, the equipment is tested for recontamination by suitable means. These suitable means may include visual examination, microbial sampling, or testing for chemical or particulate residues. Caution should be used in not just replicating the same analytical testing with the same sampling points as was done at the end of the three PQ cleaning validation runs. It is likely that any recontamination of the equipment will occur at different locations and with different residues. One verification study may be enough to establish this time limit, unless it is determined that establishment of this time limit is a critical control issue for manufacturing purposes.

A further issue related to the shelf life of cleaned equipment is what should be done if the shelf life is exceeded. If the time limit is exceeded, manufacturers have two options. One option is to repeat the cleaning procedure as per the SOP. In most cases, where the concern is particulate contamination ("dust") and the cleaning process utilizes aqueous cleaning with a surfactant, this is more than adequate. The second option involves an additional or special cleaning step. If the source of recontamination is microbial, then in addition to repeating the cleaning SOP, a sanitizing step (such as with hypochlorite or hydrogen peroxide) followed by a rinse should be included. If the sanitizing is by steam and the equipment is used for parenterals, then repeating the cleaning step after sanitizing should be considered for endotoxin reduction.

ACTION

In spray applications, *action* is related to the force with which the cleaning solution impinges on the surface to be cleaned. It is also called *impingement.* The higher the force of impingement (the higher the pressure), the more likely impingement will dislodge residues on those surfaces. It should be noted that this action only serves to dislodge residues. What happens to that dislodged residue will depend on the nature of the cleaning solution and the flow characteristics of the cleaning solution. For example, if water alone is used in a high-pressure spray application for a water soluble residue, then the dislodging of that residue may serve to increase the rate of dissolution (because of the increased surface area of the residue). On the other hand, if the residues are water insoluble, the removal of the dislodged residues will depend on the flow of water to physically remove those residues from the system. With aqueous cleaning using formulated cleaning agents (detergents), the detergent may serve to facilitate cleaning by wetting water-soluble residues (thus accelerating dissolution time) and/or by emulsifying or suspending water-insoluble residues (thus keeping them from redepositing or reaggregating within the system).

A key to performance involving this impingement action is insuring *uniformity* across all surfaces to be cleaned. In manual high-pressure spraying, this can be controlled to some extent. On the other hand, in a fixed spray device clean-in-place (CIP) application in a closed vessel, the impingement of the cleaning solution in many cases will be only on the dome of the vessel. Along the side walls and on the floor of the vessel, the contact of the cleaning solution with the surface will not involve impingement. The cleaning solution will only cascade down the side walls, which may or may not present problems. If the soil on the vessel dome is the most difficult to clean (perhaps because it is dried), then impingement due to the spray device may facilitate the cleaning process by accelerating the cleaning of the dome. On the other hand, if the nature of the soil is uniform throughout the process vessel, then the impingement force only on the vessel dome may not add any special benefit to the cleaning process (it is just a consequent feature of using a spray device in CIP processing).

It is also important to separate the effects due to impingement from the effects due to mixing or agitation (covered in Chapter 5). Impingement is the physical force of droplets or a stream of the cleaning solution impinging on a surface; agitation is movement within the cleaning solution itself. The movement within the cleaning solution can have some effect in physically dislodging soils, but the force is considerably less than that of impingement.

CONCENTRATION

The concentration of the cleaning agent in the cleaning solution is the key here. For a solvent alone (water or an organic solvent), concentration is not an issue, since the solvent is the cleaning agent and is used without dilution. However, for aqueous cleaning using detergents, the detergents are typically used at concentration from about 1 percent to 5 percent (volume/volume) in water but may be as high as 15 percent. As a general rule, the higher the concentration of the cleaning agent, the more effective the cleaning process. While this is true at concentrations of typical use, it is not necessarily the case at extremely high concentrations. For example, using a detergent at 100 percent is not necessarily more effective than cleaning at 5 percent (not to mention the rinsing issue when the detergent is used at 100 percent).

Selecting the appropriate concentration depends on a number of factors. One factor involves balancing time, temperature, and concentration depending on the objectives of cleaning. It is well known that it is possible to obtain equivalent results by cleaning at either a low concentration of cleaning agent at a higher temperature or a higher concentration of cleaning agent at a lower temperature. This same inverse relationship also generally holds between concentration and time (high concentration/short time vs. lower concentration/longer time). If a shorter cleaning time is a major goal, then a higher concentration of cleaning agent may be used. If savings of chemical costs (or perhaps disposal concerns) are of more interest, then lower concentrations for longer times will be utilized. Examples of possible tradeoffs involving relationships of the factors of time, temperature, and concentration are given in Figs 4.1, 4.2, and 4.3.

Material compatibility may also affect the selection of the detergent concentration. For example, the compatibility of caustic cleaners with glass-lined vessels may affect the selection because of concerns about long-term etching (corrosion) of the glassware. Since the inverse relationships (between temperature and concentration of cleaning agent on cleaning efficiency or between temperature and concentration of the cleaning agent on corrosion of the glass surface) are not necessarily linear, it may be possible to select a combination of concentration and temperature that

Figure 4.1. Effect of varying cleaning agent concentration and washing time.

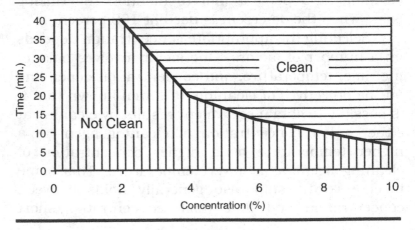

Figure 4.2. Effect of varying cleaning agent concentration and washing temperature.

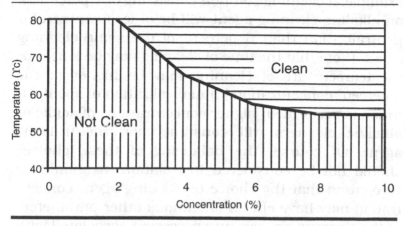

Figure 4.3. Effect of varying washing time and washing temperature.

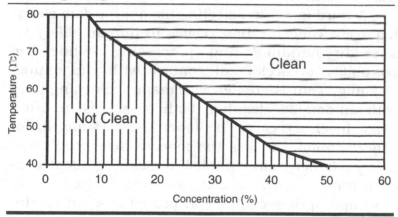

provides the same cleaning efficiency but lowers the potential for glass surface etching. This has to be decided on a case-by-case basis.

Another factor in the selection of the appropriate concentration is disposal and/or neutralization of the spent solution. The effect of an acidic or alkaline cleaning solution on the pH of a waste stream will be less with a lower cleaning agent concentration. For example, a cleaning solution consisting of 1 percent of an alkaline cleaning agent will be of less concern for neutralization than 6 percent of the same cleaning agent. (Note that the time of cleaning may be longer or the temperature of cleaning may be higher at the 1 percent concentration.) If neutralization of spent solution is performed, higher concentrations of acidic or alkaline cleaners will consume more neutralizing agent. Of course, the balancing of these choices should not be considered in isolation. It should be recognized that the choice of cleaning agent concentration may have effects on various other parameters in the cleaning process and processes associated with the cleaning process. Therefore, selection should not be considered only in light of the disposal and neutralization issue.

A third factor in selecting the appropriate cleaning agent concentration relates to human safety issues in handling and processing. While the possibility of accidental contact with the diluted cleaning agent is greater with manual cleaning than with automated cleaning, both may present problems. In either case, the lower the cleaning agent concentration, the less the concern over health and safety. Rather than establish a cleaning agent concentration based on safety, a better approach is to have appropriate equipment, procedures, and training to insure that whatever cleaning agent is used can be safely handled.

TEMPERATURE

The fourth factor to be covered in this chapter is the temperature of the cleaning solution and the temperature of the rinse. As a general rule, the higher the temperature of the cleaning solution, the more effective (faster) cleaning proceeds. As mentioned earlier, there is a "tradeoff" between such factors as temperature, time, and concentration such that relative contributions of these factors can be balanced in selecting a cleaning process to meet a given facility's objectives. One exception to this "higher is better" rule is with the cleaning of proteinaceous materials, such as in a biotechnology facility. Because the heat of the cleaning solution may help to "set" the residues on equipment surfaces, it is generally preferable to clean first with cold water. The actual cleaning (with the detergent solution) is then conducted at elevated temperatures. However, as mentioned in Chapter 3, this cleaning is preceded in many cases by a "cold" water (i.e., ambient temperature) prerinse. Another exception is in facilities where the manufacturing process is performed at low temperatures (0–10°C). In this case, there may be concerns about cycling the equipment between those low temperatures and typical cleaning temperatures of 60–80°C. Such wide variations in temperatures may cause unacceptable stress and consequent wear in the equipment. A lower cleaning temperature is therefore an acceptable compromise for equipment compatibility reasons.

Cleaning temperatures may have a strong influence on cleaning mechanisms. For example, hydrolysis of a drug active ester in an alkaline cleaning solution may not occur at ambient temperature but may proceed rapidly once an activation temperature is reached. This effect may also hold true for waxy excipients, such as those found in dermatological preparations. It is necessary to achieve a cleaning temperature close to the melting point of the waxy material in order for the material to be loosened and then emulsified by the cleaning solution. Some have proposed a rule of thumb for the relationship between cleaning time and temperature similar to the general rule for certain chemical reactions, namely that the cleaning time can be cut in half for every 10°C rise in temperature. While that rule is used for the kinetics of chemical reactions in solution (the rate of the reaction doubles for every 10°C increase), its applicability to cleaning processes is questionable. In the case of cleaning processes, physical processes such as wetting and emulsification are also involved. In addition, the cleaning process involves processes on surfaces, not in solution. Therefore, while it is generally true that cleaning is more rapid at higher temperatures, there are no good rules for correlating the relationship of time and temperature. This relationship has to be determined on a case-by-case basis.

The actual temperature of the cleaning solution may be determined by the available supply of, for example, USP Purified Water in the facility. If that water is 80°C, then cleaning is usually done at that temperature. If a higher temperature is needed, then a heat exchanger is typically used to raise the temperature. The need for temperature control during the cleaning process may also be critical. If the temperature is to be maintained at a constant temperature for the entire cleaning process, it may be necessary to have a heat exchanger in the cleaning solution circuit to maintain that temperature. The maintenance of temperature will depend on the time of cleaning as well as the degree of insulation of the process equipment and associated piping. A constant temperature for most cleaning applications would be controlled to approximately ± 5°C; for example, processing at 80°C would be controlled from 75°C to 85°C. For cleaning validation PQ runs, at least one of the runs should be deliberately planned to run at the lower end of the control range (assuming that the lower end is the worst case for cleaning).

If temperature cannot be maintained because of the length of time of cleaning, then it may be important to document the temperature change profile. A consistent cleaning process would then be defined as a cleaning process in which the temperature drops off at the same rate (within reasonable limits) for each cleaning event. In other words, if the cleaning solution *consistently* drops from 80°C to 55°C over the course of a 90-minute cleaning cycle, then the cleaning cycle could readily be validated. In that case, however, one might want to record the cleaning solution temperature at the beginning and the end of each and every cleaning cycle as a means of monitoring. If the cleaning temperature decreased from 80°C to 50°C in one run, and then from 80°C to 70°C in another run, one might question the consistency of the cleaning process.

SUMMARY

Time, action, concentration, and temperature are generally somewhat controllable and should be considered as part of the critical control parameters for the cleaning process. Time, concentration, and temperature can generally be varied widely, whereas action may be restricted within certain limits (i.e., it can be varied the least). At a minimum, the time, temperature, and cleaning agent concentration should be measured for the cleaning process to help insure consistency. Action (impingement) is difficult to measure directly, but the level of consistency may be measured indirectly by measuring spray pressure and/or flow rates. The factors covered in the next chapter are more "givens" in the cleaning process and may be controllable to a lesser extent.

REFERENCES

1.   Verghese, G. 1998. Selection of cleaning agents and parameters for cGMP processes. *Proceedings of Interphex Conference*, 17–19 March in Philadelphia, pp. 89–99.

2.   FDA. 1993. *Guide to inspections of validation of cleaning processes*. Rockville, Md., USA: Food and Drug Administration, Office of Regulatory Affairs.

5

Process Parameters in Cleaning—Part II

The remaining cleaning process parameters—surface type and quality, soil levels, soil conditions, mixing, water quality, rinsing, and environmental factors—may be "givens" in the cleaning process and can only be changed to a limited extent. However, awareness of these factors may be important for designing a rugged procedure as well as for identifying worst-case examples for cleaning validation purposes [1].

SURFACE TYPE AND QUALITY

The surface to be cleaned can affect the nature of the cleaning process. The surfaces that are being cleaned should be identified, e.g., stainless steel, glass, and a variety of plastics. Stainless steel and glass are the most common in process equipment and piping.

Plastics may be either a variety of hard plastics used on conveyor belts (for tablet filling, for example) or a variety of flexible plastics used as gaskets and seals in pharmaceutical manufacturing equipment. The main issue in terms of surfaces is whether there is any evidence that adhesion of soils to one of the surfaces is such that removal of the soil from that surface requires additional effort. If that is the case, then the focus of cleaning should be on that surface (the worst case). As a general rule, cleaning of a given soil type from glass-lined vessels is easier than cleaning the same residue from stainless steel vessels because glass is a much smoother surface (discussed below). Various plastics are also generally easier to clean unless the surface has been marred by scratches. In such cases, the crevices provide a "hiding place," making the soil more difficult to remove. Another situation that can make plastics difficult to clean is the absorption of soils into the plastic material. While the surface of the plastic may appear clean, over a period of time the absorbed material may leach out of the plastic and possibly contaminate subsequently manufactured products. The extent of leaching will depend on the chemical environment presented by the manufactured product in contact with plastic surfaces. The best approach to dealing with issues of this nature is to avoid those problem plastics where possible and select another material of construction for that item.

In addition to the type of surface, the quality or *finish* of the surface should also be addressed. As a general rule, the rougher the surface, the more difficult it is to clean. "More difficult to clean" may mean, for example, requiring a longer time to clean. For example, electropolished 320 grit stainless steel is easier to clean than electropolished 150 grit stainless steel finish. In extreme cases where the surface is physically marred, this effect may be due to the fact that crevices in the surface effectively trap the soil. An example of this may be seen in glass-lined vessels. Because of misuse or long-term use, the surface of the glass may become etched and may have a white or frosted appearance. That etching or roughness of the surface contributes to the greater difficulty of cleaning such vessels. In more moderate cases, it may be that the rougher surface provides a means for the soil to "grip" the surface. In this case, it is not unlike the painting of a smooth surface, in which the smooth surface is "roughed up" with sandpaper to enable the paint to adhere better. Parts of equipment that have rougher surfaces may be candidates for worst-case locations for swabbing purposes for residue analysis.

SOIL LEVEL

Soil level refers to the amount of soil (in units such as mg/cm^2) that may be present on different surfaces within the equipment. In most manufacturing processes, different soil levels will be present on different equipment surfaces at the time cleaning is initiated. These different soil levels may be due to differences in surface types, locations, and/or surface configurations. For example, higher soil levels can ordinarily be expected to be found on the bottom of a horizontal pipe as opposed to the top of a horizontal pipe, on any horizontal surface as compared to a similar vertical surface, on any surface involving a narrow passageway or V-shaped juncture, or on any surface where the soil can be expected to dry out.

The obvious conclusion is that the more soil present, the longer it takes to clean that surface (other things being equal). However, caution should be used in extrapolating the cleaning time solely based on the amount of soil present (in mass per surface area). If one can clean soil at the level of X mg/cm^2 with a 15-minute cleaning cycle, it is not necessarily true that the cleaning will take 30 minutes if the same soil were present at 2X mg/cm^2. On a flat surface, the cleaning time may only be 20 minutes due to the fact that the bulk of the soil is readily removed; it is only the soil near the surface itself that is more difficult to remove. On the other hand, if the same relative soil levels were present in difficult-to-clean areas (such as dead legs, etched surfaces, and crevices), it may take more than twice as long to clean at 2X mg/cm^2 because, while the "surface soil" may be more readily available, the "deeper" soil is physically less available to the cleaning agent.

Higher levels of soil may also be of concern because of the possibility of solution saturation of the cleaning agent. Finally, higher levels of soil left behind contribute to lower production yields. For extremely potent drugs, the economic value of the residues remaining should be addressed, and ways to increase yields and thereby reduce soil levels prior to cleaning should be considered.

Other things being equal, higher soil levels (per surface area) will represent the worst case for cleaning and may require a longer cleaning time (for example). It is important (but not critical) that any lab studies (see Chapter 6) done to support cleaning recommendations are performed at levels comparable to the highest levels found in the actual equipment to be cleaned. In many cases, this is not possible because lab studies are done before the process is even run on pilot-scale equipment. In these cases, it must be realized that adjustments in the scale-up process may be needed. The principle, however, is that those locations in the equipment that have the highest soil levels should at least be considered as worst-case locations for cleaning and sampling purposes. In some cases, they will be the worst cases; in other cases, they may not be the worst-case locations because of other factors. For example, a dried soil residue at a lower mass amount per surface area may represent a worse case than a significantly higher amount of freshly deposited soil.

In addition to identifying any differences between cleaning on different surfaces, it should be recognized that the juncture of two surfaces (such as around seals) can also provide differences in terms of cleaning processes. Such locations may be more difficult to clean, not because of the inherent nature of each surface but because the area at the junction permits higher levels (per surface area) of soil deposits, while also restricting access by the cleaning agent.

SOIL CONDITION

The condition of the soil at the time of cleaning may significantly affect the type of cleaning procedure required. There are four general types of soil condition: "freshly deposited" ("wet"), "dried," "baked," and "compacted." As a general rule, freshly deposited soil is the easiest to clean. A good analogy is cleaning dishes at home. If one washes the dishes as soon as the meal is complete, then the dishes are generally easiest to wash.

If the product is allowed to dry on the equipment, either because of a significant lapse between processing and cleaning or because of drying during processing, the soil may be more difficult to remove. This is certainly true of products containing such ingredients as carboxymethyl cellulose as an excipient. Using the home dishwashing analogy, letting plates sit on the table overnight will generally make them much more difficult to clean the next morning. This difficulty in cleaning may require additional detergent, a more powerful detergent (chlorinated rather than nonchlorinated), a long dishwasher cycle, or perhaps a presoak to effectively clean the soiled dishes.

Items that have baked-on soils will be even more difficult to clean. The difference between dried soils and baked soils is that baking causes some chemical change in the soil, generally making it more difficult to clean. For example, if a formulation contained a sugar solution, the sugar could be easily removed with conventional cleaning. Even if the sugar solution were dried on a surface, it could still be readily removed, perhaps with additional cleaning time. On the other hand, if the sugar were baked on and became caramelized, it may be very difficult to remove with the same cleaning process used for the fresh solution or the dried residue. A similar situation occurs in biotechnology manufacture when the manufacturing vessel is steamed before the initiation of cleaning; the steaming process "sets" the proteins on the surface, and these denatured proteins are much more difficult to clean. Here again, using the home dishwashing analogy, a pan in which a cake is baked can be more difficult to clean compared to the bowl in which the cake mix was merely prepared.

Compacted soils are those that are subject to mechanical pressure, which may change the physical nature of the deposit and may make it more difficult to remove. Compaction may retard penetration of the cleaning solution into the soil (in part by minimizing the surface area of the soil and in part by modifying the bond of the soil to the surface), thus increasing the time required for cleaning. These types of residues are most likely found in the processing of a powder. For example, as tablets are formed on a tablet press, residues on surfaces may become physically compacted. It can be expected that these surfaces will be the most difficult to clean; they may also serve as possible sources for nonuniform contamination of the next product if not adequately cleaned (see Chapter 10).

There may be some cases in which drying of the freshly deposited soil actually makes the residue easier to remove. For example, drying may produce a friable powder that could be easily blown off the surface. There may also be circumstances in which effects other than residue condition may change. For example, during an extended time before cleaning, microorganisms may proliferate, thus significantly increasing the bioburden that must be reduced by the cleaning process (prior to sanitization or sterilization).

The important issue in addressing soil condition is that a cleaning procedure to address the worst-case soil condition must be considered. In some cases, this might not be a cleaning process limiting step; however, in the majority of cases, this will be a critical factor in establishing the cleaning process conditions.

MIXING

Mixing is related to agitation. It is different from impingement in that impingement involves the force of a cleaning solution hitting a surface. Mixing involves maintaining the uniformity of the washing or rinsing solution throughout the system. This does not refer to the uniformity of the cleaning agent in solution; it is the uniformity of the soil within the cleaning solution. Without agitation of the cleaning solution (through flow, use of an impeller, or use of a spray device), it is possible that the soil to be removed will become more concentrated in the cleaning solution near the surface of the equipment. If solubility is the only cleaning mechanism (see Chapter 2), then a concentration gradient will result, with the cleaning solution nearest the surface highly concentrated in soil, and the cleaning solution farther away from the surface less concentrated in soil. This situation is counterproductive because the solution in contact with the soil (the only solution that will dissolve more soil) is the least likely to rapidly dissolve more soil. Dissolution can be maximized by mixing of the cleaning solution, so that the dissolved soil is evenly distributed throughout the cleaning solution. By minimizing any concentration gradients, the cleaning process can be shortened. A similar case would arise with other cleaning mechanisms, such as emulsification. A lack of mixing would minimize the ability of the cleaning solution to emulsify soils on the surface. In the case of emulsification, mixing may also be necessary to help maintain the emulsion. Emulsions are thermodynamically unstable, and those typically formed in a cleaning operation may have to be continuously mixed to maintain the emulsified soil. In any case, effective agitation of the cleaning solution during the cleaning process can help maximize cleaning solution performance.

WATER QUALITY

If aqueous processing is used, water quality may be critical [2]. Water quality may vary as follows:

- **Water for Injection** (WFI): meets USP (U.S. Pharmacopeia) specifications (and is essentially distilled water with low endotoxin and low bioburden).

- **Purified Water** (PW): meets USP specifications (essentially the same chemically as WFI, with higher bioburden limits and without the endotoxin specification).

- **Deionized water:** water in which ions are significantly reduced, which may be produced by a variety of methods but does not necessarily meet any USP specifications.

- **Softened water:** hardness ions (calcium and magnesium) are removed or reduced from the water.

- **Potable (tap) water:** meets only drinking water standards, and there are no restrictions on water hardness.

Issues for the washing and rinsing steps are different. For the washing step, the main issue relates to water hardness. Water hardness ions are known to affect the performance of surfactants in aqueous cleaning. If the cleaning agent contains surfactants, higher levels of cleaning agent may have to be used with tap water to achieve the same level of cleaning performance as compared to the dilution of the cleaning agent with softened (or better quality) water. Another issue with tap water is that when it is used to dilute alkaline cleaners, calcium ions may precipitate out as calcium carbonate (the carbonate being present in the water itself). This deposition of carbonate salts is not desirable and will certainly affect the appearance of the cleaned equipment. There are two options to minimize the effects of calcium carbonate deposition: (1) Use a cleaning product with chelants, such as ethylenediaminetetraacetic acid (EDTA), to tie up the calcium ions and keep them from precipitating. (2) Follow the alkaline washing step with an acid cleaner; such an acid wash will readily dissolve and remove freshly precipitated calcium carbonate. While most finished product pharmaceutical companies will use WFI or PW for the washing step, some companies have started using potable water for this step because of resource constraints. If potable water is used, water quality should be monitored on a regular basis. Depending on the water source and/or the season, water quality (such as hardness) may vary. A cleaning process should be designed to work effectively under the worst water quality conditions (e.g., highest water hardness), and at least one cleaning qualification run should be performed under those worst-case water quality conditions.

For rinsing purposes, the general practice for any pharmaceutical manufacturing is that the water quality of the *final* rinse must be at least as good as the quality of that water added for manufacturing in the subsequently manufactured product. The rationale for this is that any residues left behind due to the final rinse would be present in the water added to the next product. Therefore, there would be no special concerns about the quality of the rinse water. If aqueous cleaning is used but the subsequently manufactured products themselves are not made with water, this rule does not apply. In those cases, the effects of residues from the final rinse may have to be evaluated on different criteria (although usually in those cases, a greater concern is the removal of the water itself so that subsequent manufacture is not interfered with).

RINSING

The main issue in rinsing is that the rinse solution (whether it is water or a solvent) is capable of carrying away the soils that have been dissolved, solubilized, emulsified, or suspended [3]. Typically, if aqueous processing is used, the rinsing solution is water. As a general rule, any quality water may be suitable for the initial rinses, while (as mentioned above) the water quality of the final rinse may be dictated by the type of water used for subsequent product manufacture. One concern with potable water for the initial part of rinsing for alkaline cleaning solutions is the deposition of calcium carbonate. Even if the alkaline cleaner contains chelants, the ratio of calcium ions to chelants during the rinsing process may not be adequate to prevent calcium carbonate deposition. This is not an issue with acid cleaners used in the washing step.

Related to the issue of water quality of the initial rinse is the temperature of the initial rinse. Other things being equal, it is preferable that the water temperature during the initial stages of rinsing be at the same temperature as that of the washing solution. The rationale for this is to prevent "shocking" the system and perhaps redepositing soils from the cleaning solution. For example, if the soil has been emulsified in an 80°C cleaning solution and the rinse is with water at ambient temperature, that temperature differential may shock the emulsion and cause it to break up and redeposit the previously emulsified soil. This may be an issue only in a few cases, but it is at least a distinct possibility in certain cases.

A second issue with rinsing is an issue similar to that of cleaning: It is necessary for the rinse solution to adequately contact all surfaces for adequate removal of both the wash solution and the soils contained in the wash solution. As a general rule, if the engineering is adequate for the wash solution to contact all surfaces, then it should also be adequate for the rinse solution (providing flow rates, pressures, etc. are the same). One should be aware, however, that there may be certain cases in which particulate soils are adequately removed by the cleaning process but may subsequently block spray nozzles and cause poor rinsing coverage.

A third issue with rinsing is foam produced during the cleaning process. The foam may be due to the cleaning agent (in which case an alternative cleaning agent should be considered) or to the soil itself (proteinaceous soils can contribute to foam, not unlike the foam formed when egg whites are beaten). Rinsing of foam is difficult and may require a longer rinsing time. If not adequately rinsed, the foam will eventually collapse and leave detergent and/or soil residues on the equipment.

Rinsing conditions (primarily the volume of water and/or the rinsing time) are not easily determined in lab studies. Rinsing conditions generally must be determined by trials on scale-up or full-scale equipment. Fortunately, these are easily done as part of any prequalification cleaning trials. In trials involving once-through-to-drain rinsing, some measure of the water quality should be monitored over time. Typically for aqueous rinsing, this would be conductivity or total organic carbon (TOC). Samples of the rinse water exiting the system should be analyzed at various intervals. For example, if it is expected that rinsing should be complete in 10 minutes, an extended rinse should be performed, with TOC being monitored at 1-minute intervals between 1 and 15 minutes. The data generated can be presented graphically, as shown in Fig 5.1. What should happen is that the TOC would decrease until it levels off. This leveling off should be close to the TOC baseline of the incoming water (it will not necessarily be the same as the baseline because just passing the rinse water through a "clean" system may result in a slight increase in TOC). The time at which the TOC values level off is the time at which rinsing is essentially complete. As a safety measure, this time may be increased by a factor of, for example, 10–25 percent. It should be noted

Figure 5.1. Continuous rinsing time study: Change in rinse water parameter as function of rinsing time.

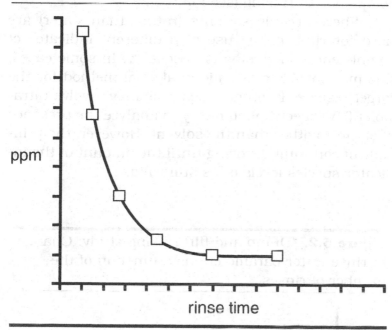

that the point at which the TOC levels off is just an indication that rinsing is complete. This leveling off does not necessarily indicate that cleaning is adequate. The completion of rinsing is merely an indication that no significant residue will be removed by further rinsing. There may, in fact, still be residues in the system that are not removed by rinsing. Analytical techniques for the determination of the adequacy of cleaning are discussed in Chapter 9. For agitated immersion systems, e.g., filling a vessel with rinse solution, agitating that solution, and then dumping it to drain, the principle is the same. However, in this case, TOC or conductivity is measured at the end of each agitation step (immediately before dumping). In like manner, the number of rinses at which the TOC (or conductivity) levels off is an indication of the practical completion of rinsing (see Fig 5.2).

Where organic solvents (rather than water) are used for rinsing, the use of a different indicator of completeness of rinsing is necessary. In some cases, this may involve a specific analytical method for the target residue. In other cases, this may involve ultraviolet (UV) spectrophotometry to analyze the presence of organics other than the solvent. However, the principle of continuing rinsing until the amount of the indicator species levels off is still valid.

Figure 5.2. "Dump and fill" rinsing study: Change in rinse water parameters as a function of the number of rinses.

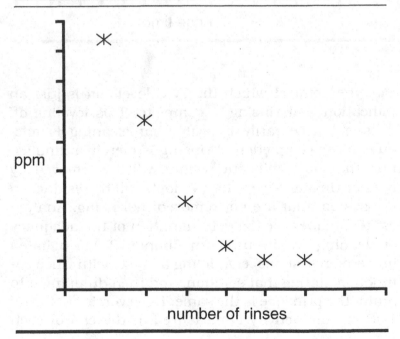

ENVIRONMENTAL FACTORS

Environmental factors such as humidity and air quality can affect the cleanliness of equipment. However, they are more related not to the actual cleaning process but to the maintenance of a clean state after cleaning. Humidity in a cleaned system may affect potential regrowth of microorganisms in a cleaned (but not sterile) system. Air quality may also affect the cleanliness in an already cleaned system by redepositing dust or particulates. These concerns should be addressed as part of the storage conditions of the equipment after cleaning but before reuse. Evidence of recontamination may require special cleaning and/or sanitizing prior to reuse.

REFERENCES

1. Verghese, G. 1998. Selection of cleaning agents and parameters for cGMP processes. *Proceedings of Interphex Conference,* 17–19 March in Philadelphia, pp. 89–99.

2. Rosen, M. J. 1978. *Surfactants and interfacial phenomena.* New York: John Wiley and Sons, Inc., pp. 281–282.

3. Lamm, E. 1990. The art of rinsing. *Precision Cleaning* 7 (1): 31–38.

6

Cleaning Cycle Development

Cycle development is the process of integrating the cleaning agents (discussed in Chapter 2), the cleaning method (discussed in Chapter 3), and the various process parameters (discussed in Chapters 4 and 5) into a usable, validatable cleaning Standard Operating Procedure (SOP). Each facility must determine the best balance for its needs within the constraints of having a validated process. While it is necessary to clean the manufacturing equipment to predetermined acceptance criteria, it is also necessary to make sure that the cleaning process is consistent and controllable and thus validatable.

VALIDATION CONCERNS

What types of things can make a cleaning process suspect from the validation point of view? One common situation is dealing with a cleaning agent that is subject to formula variation because its primary use is in other manufacturing settings. For example, products designed for the food or dairy industry may be changed (a formulation change or discontinuance and replacement by a different product) based on the needs of that industry. Where cleaning is not validated, it is much easier to substitute a cleaning agent that performs "the same" even though the chemical makeup is different. The pharmaceutical industry is not looking for new and improved products. Improvements may be the addition of a dye or perfume (something that is generally avoided in pharmaceutical manufacturing because dyes and perfumes generally add no value to the cleaning process), or they may involve a milder cleaning system (pharmaceutical manufacturers are not interested in a milder process; they want a process that is identical to the process originally validated).

Another concern with selecting a validatable cleaning process is the proper control of temperature. For manual cleaning at a sink, this may be a significant issue depending on how often the cleaning solution is prepared. The possibility of chunks of soil blocking spray nozzles in a spray ball may also raise concerns; strainers or filters to prevent this should be designed into the system as a preventive measure. Another concern could be with consistency of the cleaning equipment. For example, for manual equipment cleaning utilizing a nylon brush, the make and model of the brush should be specified to the extent possible to assure consistent results. Other considerations can also be addressed as a manufacturer addresses the issue of validatability of the cleaning process. For example, the manufacturing process of the product being cleaned should be validated or at least be considered validatable at the time at which the cleaning validation is performed. In many cases, the cleaning validation may be done on the same batches for which process manufacturing validation is performed. The rationale for this is that if the manufacturing process itself changes significantly, the nature and location of soils, and hence the cleanability of the system may vary, thus calling into question any cleaning validation work done previously.

ROBUSTNESS

In addition to addressing the issue of validatability, the pharmaceutical manufacturer should also consider the ruggedness or robustness of the cleaning process. The question that should be asked is not, "How dirty can it be and still be acceptable?", but rather, "How clean can I get it?" This approach reduces the risks of either failing the validation protocol or significantly contaminating another product. The safety margins generally designed into cleaning validation work are significant. However, once the acceptance criteria are selected, it is preferable to design a cleaning process that is not operating too near the edge of failure.

The payback from effort in a cleaning process can be illustrated by considering a typical S-curve performance, as illustrated in Figure 6.1. In this figure, the x-axis represents some measure of the cleaning effort, such as time of cleaning, temperature of cleaning, spray pressure, cleaning agent concentration. The y-axis represents some measure of how clean the system is. As cleaning effort is increased, the payback in terms of the cleanliness of the system starts to increase. At a certain point, the increase in "cleanliness" is very significant. However, as more effort is put into the system, the improvement in cleanliness

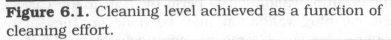

Figure 6.1. Cleaning level achieved as a function of cleaning effort.

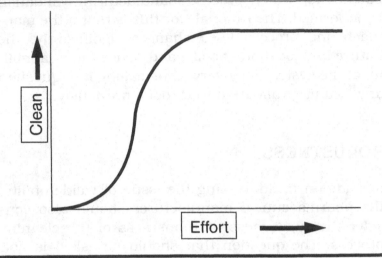

diminishes and may level off. The peak level of performance may vary depending on the cleaning process (aqueous cleaning vs. solvent cleaning, or manual cleaning vs. clean-in-place [CIP]).

If the required level of cleanliness is above the maximum possible for that system (illustrated in Figure 6.2), then it is clear that no additional effort (at least within reason) will result in an acceptably clean surface. It is not just a matter of a little longer time or a slightly higher chemical concentration. This would suggest that a different cleaning system is needed. Changes to the cleaning system to achieve acceptable results will depend on a variety of factors, including the amount of information generated as the cleaning process was evaluated. If typical performance variables such as time, temperature, and cleaning agent concentration have been evaluated, this would suggest that perhaps a different cleaning agent or a

Figure 6.2. Inadequate cleaning level achieved.

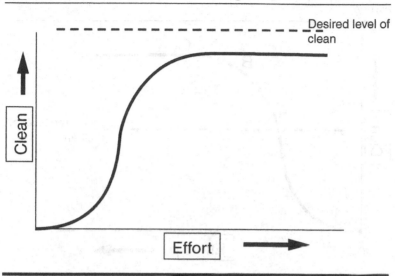

different cleaning method is required. For example, if an acidic cleaner is used and provides inadequate cleaning, perhaps an alkaline agent may work better. On the other hand, if early lab screening work suggested that better results were obtained with the acidic cleaning agent than with the alkaline cleaning agent, then perhaps a combination cleaning process (with an alkaline wash followed by a rinse, an acidic wash, and then a rinse) may provide the improvement needed. Unfortunately, when looking for a new system that might be more effective, there is no simple formula to try. The best path may be evaluating clues generated during various screening experiments.

If the required level of "clean" is below that which is achievable with the cleaning process (as illustrated by Figure 6.3), then it is necessary to decide how much "overkill" should be designed into the cleaning process. Some degree of overkill is necessary because

Figure 6.3. Selecting the appropriate level of safety margin.

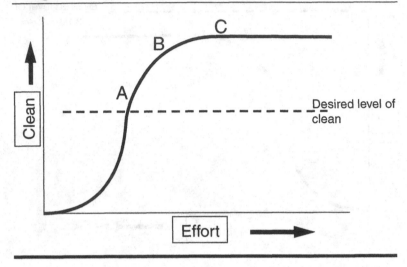

of normal variations within a cleaning process. Parameters such as temperature, time, and cleaning agent concentration are normally controlled within a defined "plus or minus" zone. There also may be normal variation in terms of items such as the amount of soil (the manufactured product) left behind to be cleaned. While it is ideal that the worst cases of all parameters are evaluated simultaneously, it is often impossible to arrange such a coincidence of worst-case parameters in three consecutive runs for validation purposes. Again referring to Figure 6.3, it would be undesirable to select a performance level at point A. It is possible that normal variations could perhaps carry the level of clean into the unacceptable region. It is more preferable to be at point B in Figure 6.3, where normal variation does not significantly affect the acceptability of the level of clean. It is most preferable to be in a region, such as that illustrated by point C in Figure 6.3, where normal variation has no effect on the level of clean. In such a case, provided that the system is performing under control, the analytical results should be such that residue levels are below the acceptance criteria with a wide margin for error. However, it should be noted that where any one company chooses to be on the performance curve is a matter of risk management, not one of scientific principle. In most cases, the costs of moving up on the curve are small compared to the costs of a failed validation effort, an investigation into cleaning deviations, or a failed lot of product.

LABORATORY EVALUATION

One of the best ways to eventually develop a successful cleaning SOP is to start with some kind of laboratory evaluation. The purpose of the laboratory study is to do the initial screening for the selection of the cleaning agent and key process conditions (time, temperature, and cleaning agent concentration) on a small scale in the laboratory. These data are then confirmed and further optimized in pilot or scale-up evaluations. This laboratory evaluation usually involves cleaning a model or simulated surface if the equipment to be cleaned is large but may involve cleaning actual parts if the items are small (and relatively inexpensive). For laboratory screening, the most typical simulated surface is a stainless steel panel with dimensions of about 3 to 20 in.^{2} (19 to 129 cm^2). Highly polished surfaces may be used but are not necessary because, in general, the rougher the surface, the more difficult it is to clean. For this reason, a stainless steel model surface may be preferred even for glass-lined vessels, because (at least in my experience) the highly polished glass surface is more readily cleaned. (Note: Etched glass may be as difficult or even more difficult to clean than stainless because of the crevices and pits in which soils can hide.)

Laboratory testing should also involve levels of soil and conditions of soil that represent the worst cases within the equipment to be cleaned. Note that the locations for the worst-case soil level and soil condition in the equipment are not necessarily the same; the soil level and condition may not even be known at the time laboratory testing is initiated. However, estimates should be made of worst-case soil levels and conditions and should be utilized for this initial screening. If there are known limitations to the cleaning process, such as the ready availability of process cleaning water at a given temperature, or the limitation that the cleaning agent is selected from those cleaning agents already approved by the facility, then this should be taken into consideration. The purpose of this lab screening is not to get the absolute best cleaning process but to help arrive at a rugged, consistent process that can be used to select a practical process to achieve acceptable cleaning performance. Therefore, as much information on practical limitations should be considered so that effort is not wasted on exploring options that could never be utilized in a given facility based on those limitations. On the other hand, it should be recognized that too many limitations may make it more difficult to achieve acceptable cleaning performance. Most likely, requiring conditions of 1 percent cleaning agent, 40°C, and a cleaning time of 5 minutes all at the same time may be unrealistic.

The method of application should approximate that to be used in the actual application; if it is difficult to simulate such cleaning in a lab study, then the method should reflect conditions that are a worse case as compared to the method to be used in actual cleaning. For example, for manual cleaning operation on small parts (or representative portions of a large part), it may be possible to closely mimic the conditions of actual cleaning. On the other hand, for a CIP application, a small CIP system may be impractical. However, in most cases, it is possible to mimic the results of a CIP operation by using an agitation immersion test apparatus. While the impingement from the spray ball is a contributing factor, it may be inconsequential for the majority of the system to be cleaned; the net effect of CIP in a laboratory study is essentially the same as that obtained under agitated immersion conditions.

Laboratory testing usually first involves testing a variety of cleaning agent types (acidic, alkaline, neutral) under conditions of high temperatures, long times, and high concentrations. The purpose of such testing is to help identify which cleaning agent is most likely to be a candidate for further screening work. Such testing may involve the use of limited parameters (such as a fixed temperature, which cannot be varied under actual cleaning conditions). The selection of cleaning agents for an initial laboratory screening process may also be determined to some extent by previous data on the cleaning of the same or similar chemical species (taking into account both the nature of the active ingredients and any excipients that may be present). The results of this first screen should be the selection of a candidate cleaning agent.

The next laboratory screen should involve the cleaning agent candidate under different conditions of time, temperature, and concentration to arrive at a combination of conditions that successfully clean the model surface in the laboratory. Here again, some of these variables may be fixed because of limitations in the manufacturing arena. It is very common, for example, for manufacturers to specify the temperature of the washing solution because of temperature limitations of available water. With such restrictions, the water temperature should be fixed, and perhaps only the time and cleaning agent concentration would be varied for evaluation purposes. It should be obvious that under conditions of fixed temperature, it may be possible to effectively clean the model surface under conditions of long times and low cleaning agent concentration or conditions of short times and high cleaning agent concentration. While it is theoretically possible to investigate numerous conditions (combinations of times and cleaning agent concentrations) that might be effective, in many cases manufacturers may have other objectives that may further limit the choices. For example, if a manufacturer requires a cleaning time of less than one hour, it makes no sense to investigate times of several hours. On the other hand, if it is not possible to produce adequate cleaning under the preliminary restrictions of time, temperature, and/or concentration selected by the manufacturer, then it will be necessary to evaluate conditions that are more rigorous than those initially selected. It should also be recognized that effective cleaning may also require a two-step cleaning procedure, such as cleaning with an alkaline detergent followed by an acidic detergent.

The test method used to evaluate the acceptability of cleaning (acceptable/nonacceptable) in the laboratory should also be selected. Certainly one standard is that the model test surface is visually clean. This may be enough for preliminary screening. Another test method employed is the so-called "water break" test [1], which can be valuable for oily residues on metal surfaces. In this test, water is allowed to cascade down the surface of the cleaned item; the presence of an oily residue on the surface causes a disruption in the smooth sheeting action of the flowing water. Weight loss can also be used and is particularly useful in initial screening involving a comparison of two conditions, neither of which produces visually clean surfaces. Specific analytical methods for the residues involved (such as an HPLC [high performance liquid chromatography] method for a particular agent) may also be utilized at this point. However, at this point in the cleaning SOP development, the analytical method may not be sufficiently developed to use as a valid screening tool. Another factor to evaluate is whether the species targeted by the analytical method is degraded during the cleaning process. If it is, then the analytical procedure may not be sufficient to determine that the surface is adequately cleaned. For most applications, a combination of visually clean and water-break free is usually adequate at this point in the cycle development.

The end result of laboratory screening should be the determination of a cleaning agent, including cleaning agent concentration, time, and temperature necessary to provide adequate cleaning in the lab-simulated cleaning situation. While there is reasonable confidence that these conditions will also work as the process is scaled up, it should be recognized that no lab screening can replicate all of the process conditions present in the scale-up equipment. The laboratory screening can only give evidence that if the cleaning agent is in contact with the soiled surfaces under the appropriate conditions of time, temperature, and cleaning agent concentration, then the surface should be adequately cleaned.

SCALE–UP EVALUATION

The first objective in any scale-up (or pilot plant) work is to confirm that the variables selected in the laboratory also work when scaled up. A second objective is to confirm that the key control parameters (such as temperature, flow rates, etc.) can be adequately controlled as the process is scaled up. For example, if a cleaning process involves 80°C water, and the cleaning time is only 10 minutes, then temperature control (with a heat exchanger) may not be critical. On the other hand, if the cleaning time were 60 minutes, then it might be expected that the temperature of the cleaning solution might decrease without the aid of a heat exchanger. Therefore, scale-up results might not be the same as those obtained in the laboratory. The reason for this is that the conditions of cleaning essentially changed (the cleaning solution was 80°C at the beginning of the cleaning cycle but had decreased to, say, 68°C by the end of the cycle).

This is only one of the many things that conceivably could go wrong during scale-up of the cleaning cycle. Any assumption made as to the quantity and nature of the soil left behind (prior to the cleaning step) must be correct. The scale-up work, therefore, is to confirm assumptions made in laboratory testing. If there are any conditions that were unanticipated, the cleaning cycle should be adjusted accordingly. If there are locations in the equipment where, for example, the drug product had been unexpectedly dried or baked onto the equipment surfaces, it can be expected that the scale-up experiments might fail to produce adequately cleaned surfaces. This may mean doing additional laboratory experiments involving dried or baked on residues, or it may mean modifications to the cleaning cycle in the scale-up equipment.

Such modifications may include lengthening the time of cleaning or increasing the concentration of the cleaning agent.

The time of scale-up work is also the time to address any optimizations of such items as time and cleaning agent concentration. The danger here is that one may be tempted to overoptimize and perhaps arrive at a point that is too near the edge of failure. The objective in most cleaning cycle development is to arrive at a rugged cleaning cycle, such that with all the normal variations (plus/minus) of temperature, time, cleaning agent concentration, and amount and nature of the residue, the results are consistently in the acceptable category. In other words, if the laboratory work demonstrated that 5 percent of a cleaning agent was effective but 4 percent was not, it is probably not worth the effort to demonstrate that 4.5 percent of the cleaning agent is also effective. It may be effective, but it is probably not worth the effort (or the risk).

One area that can be optimized during the scale-up process involves the rinsing conditions, primarily the time of rinsing (for a continuous rinsing process) or the number of rinses for discrete rinses. The procedures for this were discussed in Chapter 5. In addition to determining the rinsing time, it may also be desirable to investigate "pulse" or "burst" rinsing. Such rinsing involves taking a continuous rinse process and breaking it up into a series of discontinuous pulses or bursts. For example, if it is determined that a continuous rinse time of 10 minutes is required for adequate rinsing, it may be desirable to utilize a series of four 2.5-minute rinses, each separated by a time frame of, for example, 30 seconds. The purpose of the pulses or bursts is to prevent a rinsed system from equilibrating to produce results that apparently indicate the rinsing is completed. For example, if there is a dead leg in the system, material (consisting of soil and cleaning solution) trapped in the dead leg may drain away with a pause in the rinsing process. This is illustrated in Figure 6.4.

The evaluation at scale-up is also the time to start evaluating the sampling locations and the sampling and analytical methods. These will be discussed in more detail in Chapters 9 and 10. However, if swab sampling is to be done, locations should be chosen that are representative of the system and, more importantly, represent worst-case (or most difficult-to-clean) locations. These locations can be identified by techniques such as cleaning under suboptimal conditions (shorter times, lower concentrations) and

Figure 6.4. Pulse rinsing.

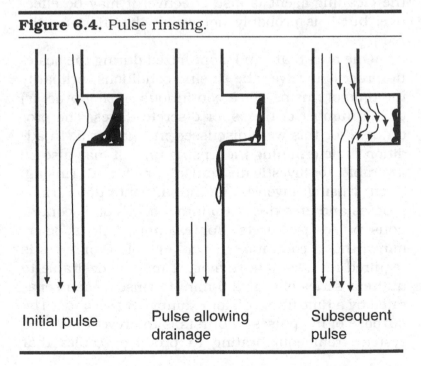

Initial pulse   Pulse allowing   Subsequent
                       drainage   pulse

observing the cleaned equipment. Since a critical element in a successful validation will be meeting the analytical limits, there should be some confirmation at the time of scale-up that these analytical limits can be achieved. Even though the analytical method and sampling techniques have not been validated with a percent recovery established, some work should be done to establish that any preliminary acceptance criteria can be met with the cleaning procedure.

The net result of the laboratory and scale-up work should be enough experimentation to have a reasonable assurance that acceptable results will be obtained when three Process Qualification (PQ) cleaning runs are conducted. The time for experimentation is during laboratory evaluation. During scale-up work, the focus is on confirmation of assumptions and tweaking of process parameters and engineering conditions. The actual validation work should not be viewed as an experiment but rather as a confirmation of what already is known to be the case. The three PQ cleaning runs are merely a documentation of what is already known to be the case.

REFERENCES

1.   ASTM F22–65. 1992. Standard test method for hydrophobic films by the water-break test. Philadelphia: American Society for Testing and Materials.

7

Grouping Strategies

"Grouping" is the concept of demonstrating that certain cleaning operations are of a similar type, and selecting one (or more) representative operations on which to conduct the three Process Qualification (PQ) cleaning runs. Under the assumptions of grouping strategies, a successful validation of the representative process means that all of the processes subsumed under that grouping are also validated. Grouping as a strategy most often applies to grouping together different products or different pieces of equipment to be cleaned. The rationale behind grouping strategies is to simplify the amount of validation work to be done based on good scientific principles and information [1,2,3,]. The Food and Drug Administration (FDA) has no formal policy on grouping (also called a "matrix" or "family" approach) for validation; however, the approach is recognized by the FDA as one that can be scientifically valid if appropriately justified [4].

The two methods of grouping for cleaning validation are "product" grouping and "equipment" grouping. In either case, products or equipment are grouped together, and then a representative case (usually the worst case) is selected for the three PQ cleaning runs. The conditions for product and equipment grouping will be considered separately.

PRODUCT GROUPING

For product grouping, the first item to consider is what products (that is, products to be cleaned) belong in the same group. For products to be grouped together, they must be similar products manufactured on the same equipment and cleaned with the same cleaning SOP (Standard Operating Procedure). Being "similar" products obviously applies to products with the same excipients and different levels of actives. However, it can also apply to products with similar excipients and different actives or even products with completely different excipients and actives. The minimum requirement is that the products are of similar types—all liquids, creams, tablets, and so on. While it is preferred that the formulations be as similar as possible, this is not an absolute requirement. However, it should be recognized that the more dissimilar the grouped products are, the more difficult it is to select a group representative.

The second criterion is that the products are all made on the same equipment. Ideally, the equipment is identical (the same piece or pieces of equipment). There is minimal risk in dealing with grouping involving multiple iterations of the same piece of equipment (for example, three powder blenders that are of identical construction). More risk is involved in similar equipment, for example, equipment of similar construction but of different sizes (this will be considered under equipment grouping considerations). However, it is probably stretching the concept of grouping too far to attempt to include processing in a glass-lined vessel and in a stainless steel vessel in one group; these should probably be validated separately.

The third criterion for grouping by products is that the products are cleaned with the same cleaning process. There is less flexibility here; every product in the group should be cleaned with the same cleaning SOP, including the method of cleaning, the cleaning agent, concentration, time, and temperature of cleaning. It is not acceptable to group together two products cleaned with the same cleaning agent but at different washing times. For example, if one product is cleaned with 5 percent of detergent X at 80°C for 60 minutes and another product is cleaned with 5 percent of detergent X at 80°C for 30 minutes, the two products should not be grouped together for cleaning validation purposes. The reason is that it becomes very difficult to select the representative product under such conditions. If cleaning time is the only difference, then the processes can be grouped together by changing the cleaning time of the second product such that the second product is also cleaned for 60 minutes. While this is clearly a case of overkill (or "overclean"), it does simplify matters so that product grouping is at least a possibility.

Once all of the products within a group have been selected (meeting the requirement of similar products manufactured on the same equipment and cleaned with the same cleaning SOP), the next step is to select a representative product. The goal here should be to select the product that is the worst case or most difficult to clean. If the most difficult-to-clean product can be acceptably cleaned, then all the other products in that group should also be effectively cleaned.

There are basically four types of justifications used to select the worst case among a group of similar products: (1) lab simulation cleaning studies, (2) solubility characteristics of the drug substance, (3) solids or activity level of the drug substance, and (4) operational experience in previous manufacturing. Each of these will be covered briefly.

Lab studies usually involve application of the different products to a model hard surface (e.g., stainless steel) followed by an evaluation of cleaning performance under different conditions (e.g., time, temperature, cleaning agent concentration) that would allow one to make a relative comparison of the ease of cleanability of the products. This evaluation should be done with the cleaning agent that is intended to be used in the manufacturing process and typically should involve an evaluation of cleaning performance at shorter times, lower temperatures, or lower cleaning agent concentrations. Evaluation under these less severe conditions may separate the products according to their ease of cleaning. The product that requires the most severe conditions to achieve acceptable cleaning results can be considered the most difficult to clean or the worst case for that product group. Note that the worst-case selection is specific to a cleaning process. Within a product group, the worst-case soil (product) using cleaning agent A is not necessarily the worst case using cleaning agent B. Of course, one of the criteria for products to be grouped together is that they are cleaned with the same cleaning SOP.

A second type of justification involves comparing of the solubilities of the drug substances in the group.

The product with the lowest solubility is considered the worst case. A comparison of solubilities should be done with the solvent used for cleaning. If an aqueous cleaner is used, then the comparison of solubilities should be at the same pH as the aqueous cleaning agent used. For example, if an alkaline cleaning agent is used, then the best comparison would be solubilities of the drug substances at the elevated pH of the cleaning agent. This should be considered because the solubilities of drug substances may change considerably with pH. In a group of products to be compared, drug A may be least soluble at pH 7, drug B may be least soluble at pH 12, and drug C may be least soluble at pH 3. Fortunately, measuring solubilities of drug substances at different pH's is a relatively easy task. This determination is applicable for finished drugs only if the excipients in the drug products being grouped are the same or very similar. This consideration for finished drugs is very important because it is often the excipients in a formulation that make it more difficult to clean. If the excipients are the same, then the different drug actives can be compared by solubility to determine the representative worst case for cleaning validation grouping.

A comparison of solids/activity is used for drug products containing the same actives but at different active levels. For example, a liquid drug may be sold with active levels of 200 mg/L, 350 mg/L, and 500 mg/L. In this case, the highest actives level would be chosen as the worst case for product grouping. It should be noted that there may be some issues here, particularly if the excipients are the most difficult to clean components of the formulation and if the active is readily water soluble. A case could be made that it is actually the formulation with the lowest actives that is more difficult to clean, since a higher amount of the more readily water soluble active may cause the residue to be more easily dissipated. However, this is probably splitting hairs; in most cases, manufacturers would use the highest actives level. However, the most conservative grouping position would be to validate separately both the highest and lowest actives level.

Operational experience can be used if there is past experience with cleaning the various products, either in a manufacturing setting or in a pilot/scale-up facility. Care should be taken that comparisons of the relative cleanability of different products are based on cleaning results with the same or similar cleaning processes. If this approach is taken, the justification needs to be documented in a report written by a scientist qualified to make that comparison.

For any of the four options used to justify the selection of the worst-case product for a product grouping strategy, clear documentation, including the rationale for the justification process, should accompany the recommendation for selection of the worst-case product. It should be noted that in some rare cases it may be difficult to select the worst-case representative product. If none of these criteria produce a product that clearly should be the representative product for grouping purposes, the manufacturer has two options: either don't group or group with the arbitrary selection of one of several equally representative products. If the latter option is chosen, then it should be clearly documented why it was concluded that no one product was singled out as the most difficult to clean. The representative product should then be selected as the product with the lowest analytical acceptance limit.

The final consideration for product grouping is the analytical limit to be set as the acceptance criterion for the three cleaning validation PQ runs. In general, the acceptance limit for the representative product should not necessarily be the acceptance limit calculated for that representative product but rather the lowest limit among all the products in that category. If the representative product has the lowest limit, then validation is straightforward in that the limit of that representative product is used. To determine the lowest calculated residue limit among the product group, it is necessary to calculate the residue limit per surface area for each product in the group (see Chapter 8 for more on residue calculations). An example is given in Table 7.1.

The representative product in this case (the most difficult to clean) does not have the lowest calculated acceptance limit. In this case, product 2 has the lowest limit. Therefore, the preferred option is to perform the three PQ cleaning runs with the representative product but measure the active A not to a limit of 3.5 μg/cm^2 but to the lowest limit in the group (1.5 μg/cm^2 of A). The logic here is that if one can clean the group representative to a limit of 1.5 μg/cm^2 of A, then one should also be able to clean product 2, which is easier to clean, to a limit of 1.5 μg/cm^2 of B. An alternative sometimes used is to clean the representative product to its acceptance criteria, and in

Table 7.1. Residue Limit Determination for Representative Product in Grouping Strategy

Product	Active Species	Calculated Residue Limits
Group representative	A	3.5 μg/cm^2 of A
Product 2	B	1.5 μg/cm^2 of B
Product 3	C	2.5 μg/cm^2 of C

three separate PQ runs, evaluate the cleaning of the product with the lowest analytical limit (product 2 in this case) at its acceptance limit. This is usually done for business reasons (to minimize risks) and does not necessarily have a valid scientific basis.

There may be cases where it is possible to clean the group representative to its analytical limit, but because of limitations in the analytical method, it may not be possible to measure the active in the group representative at levels of the lowest acceptance limit in the group, e.g., due to the limits of quantitation of the analytical method. One option in this case is to break up the group and remove the product with the lowest limit from the group, validating it separately. Then the group is validated using the representative product utilizing what is now the lowest limit among the products remaining in the group. A second (and riskier) option is to clean the representative product to its acceptance limit and then perform at least one qualifying cleaning run on each of the other products in the group using the analytical acceptance criteria for each product tested. In the example given in Table 7.1, three PQ runs would be performed with the group representative, with the acceptance criteria being 3.5 $\mu g/cm^2$. Then separate qualifying runs would be performed with product 2 using an acceptance criterion of 1.5 $\mu g/cm^2$ of active B and with product 3 using an acceptance criterion of 2.5 $\mu g/cm^2$ of active C. Of course, this extra work in the qualifying runs minimizes the impact of grouping strategies somewhat. The value of a grouping strategy in such a case has to be evaluated on a case-by-case basis. Conditions under which product grouping strategies are permitted in an individual facility should be detailed in the cleaning validation master plan for that facility.

EQUIPMENT GROUPING

The idea behind equipment grouping is classifying different pieces of equipment together, selecting a group representative, and then performing the three PQ cleaning runs on that one piece of equipment. Care must be used in grouping equipment together. If equipment is to be grouped, it must be equipment cleaned with the same cleaning SOP. It is generally unacceptable to group different types of equipment together. For example, it is generally unacceptable to group together a ribbon blender and a V-blender. The main reason for this is that even though the chemistry of cleaning may be the same (same cleaning agent, concentration, time, and temperature), the engineering aspects of cleaning are sufficiently different to require separate validation. If equipment grouping is performed, it is most commonly done on pieces of equipment with a similar design having either the same or different sizes. Most commonly, this would be done on equipment such as storage vessels. For example, stainless steel storage vessels of the same design that are 300 L, 500 L, and 1,000 L in size could be grouped together for cleaning validation purposes.

Once an equipment grouping is established, the next issue is to select the representative piece of equipment in that group for the three PQ cleaning runs. Unfortunately, this is more difficult than selecting the representative product for product grouping. Is there any evidence that a 300 L storage tank is any more difficult to clean than a 1,000 L storage tank of the same design? If there is such evidence, then that evidence should be documented as justification for selecting that equipment size as the representative for PQ runs. If there is no evidence suggesting why one size would be more difficult to clean, then there are two options for grouping purposes. One is to conduct three PQ runs on the equipment group, making sure that at least one PQ run involves the largest size and at least one PQ run involves the smallest size. In the case of the three storage vessels, one PQ run should be performed on the 300 L vessel and one on the 1,000 L vessel, with the third PQ run being performed on any one of the three sizes. The assumption here is that with at least one of the PQ runs at either extreme, all sizes could be validated together. A second option is to perform three PQ runs on the largest size and three PQ runs on the smallest size. This option involves separately validating the largest and smallest sizes and assuming that any size in between would also be covered by that validation.

A special case of equipment grouping involves smaller items that are cleaned manually or in an automated parts washer. Must all of these items be validated separately or is it possible to group them together for validation purposes? As with the grouping of larger process vessels, the key to grouping smaller parts is that they are cleaned with the same cleaning SOP. If some parts are cleaned by a manual sink scrub and others are cleaned with high pressure spray cleaning, then all of the items cannot be grouped together. Once those items in the equipment group are established, the next step is to identify those items that are most difficult to clean. Note that in this case, dissimilar items can be grouped together (provided they are cleaned with the same cleaning SOP and have the same soil or residue type). In other words, simple items (which may be relatively easy to clean) can be grouped with more complex items (which because of, for example, their geometry, can be more difficult to clean). The validation process would then include justification of the selection of those parts deemed most difficult to clean and PQ runs on those items. If the items that are most difficult to clean are successfully cleaned, then the items that are easier to clean should also be effectively cleaned.

There are some additional concerns that should be addressed in washing items in an automated parts washer, including the loading of the parts washer. (Any worst-case loading, in terms of loading pattern, should be identified and considered for validation.) Also, any information relating to worst-case locations within the washing chamber should be addressed in terms of selecting the worst-case locations for sampling purposes.

GROUPING BY CLEANING PROCESS

One option that must be rejected is grouping together products or equipment cleaned by different cleaning processes. Such an attempt at grouping is not defendable for cleaning validation purposes. For example, if product A is cleaned for 60 minutes, but product B using the same cleaning solution requires only 30 minutes, then it would be difficult to group the two products together and have a rational basis for selecting the worst-case process. In such a case, the preferred technique to use is to default from the shorter cleaning time (30 minutes) to the longer cleaning time (60 minutes) for product B. Under such conditions, it becomes possible to group the two products (they both now utilize the same cleaning SOP) as well as to select the worst-case product to clean (product A, because it requires a longer time to achieve acceptable cleaning).

Another case that may present itself involves two products, one cleaned with alkaline cleaning agent C and another product effectively cleaned by acidic cleaning agent D. Here again, the two products cannot be grouped together because two different cleaning SOPs are used. One way to effectively group these two products for cleaning validation purposes is to modify the cleaning SOP such that it involves cleaning first with the alkaline cleaning agent, rinsing, cleaning with the acidic cleaning agent, and rinsing. The rationale behind this is that now there is one cleaning SOP for both products. The alkaline cleaning solution should effectively remove the first product. The subsequent acidic cleaning step essentially involves cleaning an already-cleaned vessel, so the overall process should be effective for cleaning the first product. The only concern with the second product is whether or not the imposition of the alkaline cleaning step has an effect on residues such that those modified residues are no longer effectively cleaned by the acid cleaning process. Although rare, this interaction should be evaluated with lab screening prior to proceeding with any grouping strategy based on the joining of two cleaning processes (such as a combination cleaning process mentioned in Chapter 2).

The introduction of a new product to be cleaned on equipment for which there is existing cleaning validation using a grouping strategy should be carefully justified and documented. Conditions under which grouping strategies can be used within a facility should be documented in the cleaning validation master plan for that facility. Justification of decisions as to grouping and the selection of worst cases should be carefully documented.

REFERENCES

1. PDA. 1998. *Points to consider in cleaning validation.* PDA Technical Report 29. Bethesda, Md., USA: Parenteral Drug Association.

2. Jenkins, K. M., and A. J. Vanderweilen. 1994. Cleaning validation: An overall perspective. *Pharmaceutical Technology* 18 (4): 60–73.

3. Walker, A. J., and M. V. Mullen. 1995. How to establish, evaluate, and maintain a cleaning validation program. Paper presented at Pharm Tech '95, 18 Sept. in East Brunswick, N.J.

4. FDA. 1999. *Human drug cGMP notes,* vol. 7, no. 1. Rockville, Md., USA: Food and Drug Administration, Center for Drug Evaluation and Research.

8

Setting Acceptance Criteria

An important consideration in cleaning validation is the determination of "how clean is clean enough?" The Food and Drug Administration (FDA) does not establish analytical acceptance criteria for manufacturers [1]. Specific analytical acceptance criteria for target residues must be established by the manufacturer. It is important in setting acceptance criteria that the limits are scientifically justified. An arbitrary setting of limits is just that—"arbitrary"—and may raise concerns in any regulatory investigation. This situation is somewhat complicated by the fact that sometimes the term *limit* is used loosely, referring to the acceptance limit in the next product, of surface contamination, or of the analyzed sample. While all these are interrelated, they are not necessarily of the same units or magnitude. For example, in the contamination of the next product (the product subsequently manufactured in the cleaned equipment), the units typically are ppm or $\mu g/g$; for surface contamination, the units are usually $\mu g/cm^2$; for the analyzed sample, the units are typically μg or $\mu g/g$. It should be clear that limits per surface area are different units and cannot be compared directly without other pieces of information (such as batch size and equipment surface area). In addition, limits in the subsequent product may be the same units as limits in the analyzed sample, but they also are not comparable without other information such as area swabbed and the swab recovery factor. What this means is that an acceptance limit of 3.2 ppm in the subsequent product is not necessarily the same as 3.2 ppm in the analyzed sampled prepared by a swabbing procedure. It is important to make these proper distinctions when discussing residue limits. This will insure that analytical methods are properly validated considering the appropriate limits for the residue in the analyzed sample. In the subsequent discussion, the calculation of residue limits for the different parts of a validated system will be discussed. The limitations and applicability of such calculations as applied to finished drug products will then be explored.

One of the main objectives of the cleaning process in drug manufacture is to remove residues of the just-manufactured product so that those residues are not transferred to the subsequently manufactured product. A key complicating feature of cleaning is that it involves not only the product being cleaned but also the product subsequently manufactured in the cleaned equipment. The starting point for any determination of residue acceptance limits is the amount of residue from the cleaning process that could be present in the subsequently manufactured product without posing an unreasonable risk. Clearly, one would prefer that no residue is present. However, it is impossible to measure "no residue." Even the criterion of being below the limit of detection (LOD) of the analytical procedure is not by itself a very good method for selecting residue limits. As methods are improved and have even lower LODs, a cleaning process which was previously viewed as acceptable can become unacceptable. The FDA's guidance for determining residue limits is that they must be logical (based on an understanding of the process), practical, achievable, and verifiable [2]. Fortunately for those involved in cleaning validation, the FDA mentions limits proposed by industry representatives, such as 10 ppm, biological activity levels such as 1/1,000 of the normal therapeutic dose, and organoleptic levels such as no visible residue. This is clearly a reference to work done by Fourman and Mullen [3] at Eli Lilly (also listed in the reference section of the FDA guidance document). While not officially "blessed" by the FDA, this Lilly method of establishing residue limits (or some variation of it based on the same principles) is widely used within the pharmaceutical industry for determining acceptable levels of chemical residues [4,5,6,7].

The published Lilly criteria are that (a) the equipment is visually clean, (b) any active agent is present in a subsequently produced product at maximum levels of 10 ppm, and (c) any active agent is present in a subsequently produced product at maximum levels of 1/1,000 of the minimum daily dose of the active agent in a maximum daily dose of the subsequent product. While Fourman and Mullen directly calculate the surface area contamination based on these latter two criteria, the analysis in this book will use the same assumptions to arrive separately at subsequent product, surface area, and analytical sample residue limits. Such a discrete analysis may be more conducive to understanding contributions to residue limits from various factors.

LIMIT IN SUBSEQUENT PRODUCT

For the calculation of the limit of the active agent in any subsequently manufactured product, the information needed is the minimum daily dose of the active being cleaned and the maximum daily dose of the subsequently manufactured drug product. For illustration purposes, product A will refer to the product being cleaned, and product B will refer to the subsequently produced product. Such a limit (L1) can be expressed as follows:

$$L1 = \frac{(0.001)(\text{min. daily dose of active in Product A})}{\text{max. daily dose of Product B}} \qquad (1)$$

For an example involving orally dosed liquids, assume that product A has an active at a level of 2,000 μg/mL and is dosed at 5 mL from 3 to 5 times daily. The minimum daily dose is

$$\left(2,000\frac{\mu g}{mL}\right)\left(5\frac{mL}{dose}\right)\left(3\frac{doses}{day}\right) = 30,000\frac{\mu g}{day}$$

If it is also assumed that product B is dosed at 5 mL from 2 to 4 times a day, then the maximum daily dose of Product B would be

$$\left(5\frac{mL}{dose}\right)\left(4\frac{doses}{day}\right) = 20\frac{mL}{day}$$

Note that the calculation for the subsequent product is independent of what the active is or at what level that active is present. The residue limit in the subsequent product can be calculated by substituting these values into Equation 1 as follows:

$$L1 = \frac{(0.001)(30,000 \ \mu g/day)}{20 \ mL/day} = 1.5 \ \mu g/mL$$

This calculation of 1.5 μg/mL (or approximately 1.5 ppm assuming a specific gravity of 1.0) is independent of batch size and the surface area of the equipment. This means that this calculation can be done as soon as information on the composition and relevant dosing of the two products is available. The calculated value of 1.5 ppm should be compared to the 10 ppm "default" value (from Lilly's criterion b) and the lower value used for subsequent calculations. It should be noted that the 10 ppm default is strictly an arbitrary value and is difficult to justify scientifically. However, if applied correctly (10 ppm is used only if it is lower than the L1 limit), it results in a limit being set that is lower than what is scientifically calculated. Therefore, since it results in an even lower limit than the L1 limit, it cannot be logically rejected as an invalid tool. For the example used, 1.5 ppm would be used for subsequent calculations (since 1.5 ppm is less than 10 ppm).

Safety factors other than 0.001 could be selected. For example, safety factors of 0.001 for oral dose product and 0.0001 for parenterals have been suggested [8]. While more stringent safety factors may be easily justified, it would require significant justification if a safety factor less stringent than 0.001 were used. However, different safety factors may be appropriate if they are applied to dosing factors other than the minimum daily pharmacological dose. In addition to various safety factors, companies may also base limits on parameters other than therapeutic doses.

For example, rather than using the minimum daily dose of the active, other measures such as the no observable effect level (NOEL) or the minimum pharmacological effect level may be selected. Since these will result in residue limits more stringent than limits based on the minimum daily dose, there is no scientific reason not to select these criteria. It should be noted, however, that arbitrarily selecting more stringent criteria may result in unreasonable cleaning overkill and may also stretch the detection limits of available analytical methods.

LIMIT PER SURFACE AREA

Once the residue limit in the subsequent product is determined (using the lower of Lilly criteria b and c above), the next step is to determine the residue limit in terms of the active ingredient contamination level per surface area of equipment. This limit (L2, in $\mu g/cm^2$) depends on the limit in the subsequent product (the lower of L1 and 10 ppm), the batch size of the product B (in kg), and the shared equipment surface area (in cm^2). This is expressed mathematically as

$$L2 = \frac{(L1)(\text{batch size of subsequent product})(1,000)}{\text{shared equipment surface area}} \quad (2)$$

where 1,000 is a conversion factor to account for ppm (limit in subsequent batch) and to convert kg to μg (for the batch size of subsequent product). Continuing forward with the example used, the limit in the subsequent product is 1.5 ppm. Assuming the batch size of the subsequent product is 200 kg and the shared equipment surface area is 60,000 cm^2, then the L2 limit is

$$L2 = \frac{(1.5 \text{ ppm})(200 \text{ kg})(1,000)}{60,000 \text{ cm}^2} = 5.0 \ \mu g/cm^2$$

In determining the surface area, all shared product contact surfaces, including piping, baffles, and the like, should be considered. If estimates are to be made, these estimates should be on the high side for surface area because larger surface areas result in lower L2 residue limits.

It should be noted that this calculation for L2 assumes that the residue will be evenly distributed over all surfaces. In fact, this is generally not the case. However, this assumption is still the worst case. If one is doing swab sampling on the most difficult-to-clean locations, then assuming an even distribution will make it more difficult to meet the residue limits for those locations. In addition, it would be difficult to scientifically justify one area of the equipment having a residue limit of 2 $\mu g/cm^2$ with other areas having limits of 0.5 $\mu g/cm^2$. If a true "sampling rinse" is used, then issues related to uneven distribution don't arise, because one is sampling the entire equipment surface [9].

If more than one product (for example, products B, C, D, and E) could possibly be manufactured following product A, then the surface area limits (L2) for cleaning following product A should be calculated for each subsequent product. The residue limit for cleaning validation purposes should be set at the lowest of these L2 surface area limits. This gives the manufacturing department more flexibility to make products in any order. It should be recognized, however, that there may be circumstances in which residue limits may require restrictions on which products may follow product A on the manufacturing schedule.

LIMIT IN THE ANALYZED SAMPLE

While one establishes residue limits for the active in the subsequent product (L1) and for the active per surface area following cleaning (L2), these are not typically *directly* measured by the analytical procedure. The analytical procedure typically measures the active in solution as a result of either swabbing and desorbing that swab into a suitable solvent or by doing rinse sampling and measuring the active in the rinse solvent [10]. For purposes of expediency, the focus here is on swab sampling. The reader is referred to other sources for examples related to rinse sampling [9].

For swab sampling, it is assumed that a specified surface area of the equipment is sampled and that the swab is then desorbed into a fixed amount of solvent. To determine the residue limit (L3 in μg/g or ppm) in the analytical sample (the solvent the swab is desorbed into), one must know the surface area residue limit (L2), the surface area swabbed (in cm^2), and the amount of solvent the swab is desorbed into (in g). The limit in the analyzed sample is calculated as follows:

$$L3 = \frac{(L2)(\text{swabbed surfaced area})}{\text{amount desorption solvent}} \qquad (3)$$

Continuing with the example used so far, if L2 is 5.0 μg/cm^2 and assuming the surface area swabbed is 25 cm^2 and the amount of solvent used for desorption is 20 g, then the limit L3 in μg/g is

$$L3 = \frac{(5.0\ \mu g/cm^2)(25\ cm^2)}{20\ g} = 6.3\ \mu g/g\ (\text{or } 6.3\ ppm)$$

RECOVERY FACTORS

As discussed in the above example, 6.3 ppm is the acceptance limit in the analyzed sample. It should be noted that this value should be adjusted by a swab recovery factor. There are two ways to do this. One is to include the swab recovery factor in the actual analytical calculation. For example, if the swab recovery factor was 0.80 (80 percent), and one measured 1.3 ppm in the analytical procedure, then that value is adjusted by dividing the analytical result by the recovery factor to arrive at a determination of 1.3 ppm/0.80 = 1.6 ppm. The other alternative is to include the recovery factor in the numerator of Equation 3 above. In this case, the recovery factor of 0.80 should be included in the numerator. While the numbers used will be different, the net effect of comparing the analytical result to the calculated limit will be logically the same. One should standardize how this is performed in order to avoid situations in which the recovery factor is used in both the calculation of the L3 limit and the determination of the analytical result on the desorbed solvent.

EFFECTS ON ANALYTICAL
METHOD VALIDATION

It is important to point out that, at least in this example, the limit in the subsequent product (L1) was significantly different from the limit in the analyzed sample (L3), 1.5 ppm and 6.3 ppm respectively. The reason for this is that L1 reflects the residue of the active being evenly distributed in a batch of the subsequent product, whereas L3 reflects the residue of the active being concentrated in what for most cases is a smaller volume of matrix material (the desorbing solvent). One can see that this effect can be leveraged by either sampling a larger surface area or by desorbing the swab into a smaller amount of solvent. In sampling a larger area, one needs to consider whether the sampling of a larger area might also lower the swab recovery factor and thus add more uncertainty to the determination. This is significant for analytical method purposes because the analytical method chosen for determining the residue of the active agent should be validated (at least in the example used) not based on the L1 limit of 1.5 ppm but rather on the L3 limit of 6.3 ppm (adjusted appropriately by the recovery factor). In this case, one might validate the analytical method not in the range of 0.5 to 1.5 ppm but rather in the range of about 2.1 to 6.3 ppm. This fourfold factor in limit of quantitation (LOQ) may be significant for analytical method selection and validation.

VISUAL CLEANNESS

The issue of visual cleanness is significant. The point is that if a surface is visually dirty, then either the cleaning procedure is not acceptable or a once acceptable procedure is now out of control. The standard of "visually clean" can be used for both validation and monitoring purposes. The dividing line between visually clean and visually dirty is usually regarded as being in the range of 4 μg/cm^2 [3]. If the L2 surface contamination acceptance limit is calculated and found to be significantly above 4 μg/cm^2, then provided that critical surfaces are readily visible, it may be possible to default to visually clean as the only acceptance criteria. For potent drugs where the L2 acceptance criterion would typically be well below 1 $\mu g/cm^2$, a determination of visual cleanliness would have no significance as to the adequacy of cleaning. In this case, visually dirty would still be an indication of cleaning failure, but visually clean could not clearly indicate whether the residue was at an acceptable level. For cases where determination of visually clean is critical, it may be appropriate to actually determine the highest level that is not visible for that specific residue. This can be done by spiking model surfaces (for example, stainless steel coupons) with different levels of the residue (for example, 0.5, 1.0, 2.0, 4.0, and 8.0 $\mu g/cm^2$) and having a trained panel of observers look at the coupons in a "blinded" manner to determine whether or not the coupons are visually clean. This should be done under viewing conditions (lighting, angle, distance) that simulate the viewing of actual equipment. The highest residue level at which all panel members consider the coupons visually clean establishes an acceptance level for that particular residue. It should be noted that in at least one document, the FDA has stated that relying on visual examination alone is not scientifically sound [1]. It is unclear from the context whether a calculation that the visual level would be less than a L2 limit would be sufficient additional information that would make the use of visual examination (without additional swab or rinse testing) appropriate.

NONUNIFORM CONTAMINATION

The cleaning validation acceptance criteria used by Lilly and referred to in the FDA guidance document on cleaning validation do provide one logical construct for determining residue limits for chemical residues of active ingredients in finished drug manufacture. One issue that may arise involves possible nonuniform contamination of the subsequent product [11]. It should be noted that this is different from (although it may be related to) nonuniform contamination of the cleaned equipment. Nonuniform contamination of the subsequent product is more likely to arise with continuous processes rather than batch processes. For example, in a batch process of liquid drug blending, the contaminating residue (provided there is adequate mixing) is likely to be evenly distributed throughout the subsequently manufactured product. In a continuous process, such as flow through a piping system or a filling nozzle in a packaging operation, it is more likely that the contaminating residue would appear in the first vials filled. This, of course, depends on the solubility of the residue in the subsequently manufactured product and on product flow characteristics. A special case of a continuous process is a tablet press, in which any residues on the tablet press surfaces are not likely to be evenly distributed over all tablets manufactured in a batch.

In dealing with nonuniform contamination, at least two alternatives are possible. If it is reasonable that the contaminating residue would appear in the first part of a product batch (of filled vials, for example), then calculations can be made to determine what would be the maximum number of vials filled that could theoretically be at the maximum allowable contamination level. Those vials, and as a safety measure a reasonable number of subsequently produced vials, should be discarded or destroyed. For example, suppose the acceptance limit for the target residue is 5 ppm in the subsequently manufactured product, which is filled as 5 mL vials. If the calculated residue levels in the filling equipment and associated piping are 3.0 µg/cm^2 (determined from a swabbing procedure, for example), and if the surface area of the filling equipment and associated piping is 5,000 cm^2, then if all the contamination in the equipment were concentrated in the first vial produced, the concentration of residue in that one vial would be

$$\frac{(0.3 \ \mu g/cm^2)(5,000 \ cm^2)}{5 \ mL} = 300 \ \mu g/mL \ (300 \ ppm)$$

This would clearly be above the L1 acceptance limit of 5 ppm. If all the contaminating residue in the equipment were evenly divided in the first 60 vials, then each vial would contain 5 ppm of residue. In that case, those 60 vials, plus a reasonable number of subsequently produced vials, should be discarded.

These determinations were based on theoretical considerations of what could happen and represent a worst case. If the number of vials to be discarded is unacceptably high, then a study could be done to deal with nonuniform contamination of the filled vial in this example. This study would involve cleaning the filling equipment and then filling a placebo product. Vials 1, 10, 20, and so forth would then be analyzed for the target residue. If vial 10 was measured at 6 ppm and vial 20 was at 2 ppm, then those first 20 vials, plus as a safety factor a reasonable number of subsequently produced vials, should be discarded.

MICROBIOLOGICAL CONTAMINATION

A second issue in setting limits is what to do about microbiological contamination. Setting acceptable limits for microbiological contamination is a more difficult issue. The main issue with microbiological contamination is that merely one organism in the equipment could possibly result in a significantly higher contamination level in products manufactured in that equipment. No clear guidelines for process equipment exist. This may be the reason why the FDA guidance document on cleaning validation explicitly states that the guidance document applies only to chemical residues. One cannot expect the equipment to be free of all microorganisms, especially if any final rinse involves nonsterile water, unless a final sanitizing or sterilization step is used. As a minimum, the criteria used for critical cleanroom surfaces should be used [12]. A second concern is with the species of microorganism present. Obviously, the presence of enteric organisms such as *Escherichia coli* or *Enterococcus* would ordinarily be unacceptable. Microbial issues are covered in more detail in Chapter 11.

Other concerns include setting limits for residues from chemical sources other than active ingredients (such as from excipients or cleaning agents), setting limits for residues in active pharmaceutical ingredients, and accounting for residues from multiple process steps. The consideration of residue limits in these cases should be based on the same principles discussed here. The first consideration is the possible effect of any residue when it is present in any subsequently manufactured product. It is then possible to work backwards to determine limits on equipment surfaces and/or bulk actives. Any residue limit determination should be based on similar considerations and principles utilized for finished drug products, modified as they apply to those situations. The key, as in any validation activity, should be sound scientific and logical reasoning.

REFERENCES

1. FDA. 1998. *Human drug cGMP notes*, vol. 6, no. 2. Rockville, Md., USA: Food and Drug Administration, Center for Drug Evaluation and Research.

2. FDA. 1993. *Guide to inspections of validation of cleaning processes*. Rockville, Md., USA: Food and Drug Administration, Office of Regulatory Affairs.

3. Fourman, G. L., and M. V. Mullen. 1993. Determining cleaning validation acceptance limits for pharmaceutical manufacturing operations. *Pharmaceutical Technology* 17 (4): 54–60.

4. Vitale, K. M. 1005. Cleaning validation acceptance criteria. Paper presented at Cleaning Validation PharmTech Conference '95, 17 Sept. in E. Brunswick, N.J.

5. Brewer, R. 1996. Regulatory aspects of cleaning validation. Paper presented at ISPE seminar, 6–8 March in Rockville, Md.

6. PDA Biotechnology Cleaning Validation Subcommittee. 1996. *Cleaning and cleaning validation: A biotechnology perspective*. Bethesda, Md., USA: Parenteral Drug Association.

7. LeBlanc, D. A. 1998. Establishing scientifically justified acceptance criteria for cleaning validation of finished drug products. *Pharmaceutical Technology* 22 (10): 136–148.

8. Hall, W. A. 1997. Cleaning for bulk pharmaceuticals chemicals (BPCs). In *Validation of bulk pharmaceutical chemicals*, edited by D. Harpaz and I. R. Barry. Buffalo Grove, Ill., USA: Interpharm Press. pp. 335–370.

9. LeBlanc, D. A. 1998. Rinse sampling for cleaning validation studies. *Pharmaceutical Technology* 22 (5):66–74.

10. Kirsch, R. B. 1998. Validation of analytical methods used in pharmaceutical cleaning assessment and validation. In *1998 analytical validation in the pharmaceutical industry*, supplement to *Pharmaceutical Technology*, pp. 40–46.

11. Jenkins, K. M., and A. J. Vanderweilen. 1994. Cleaning validation: An overall perspective. *Pharmaceutical Technology* 18 (4):60–73.

12. Pharmacopeial Forum. 1997. Microbial evaluation of clean rooms and other controlled environments, in-process revision. *Pharmacopoeial Forum* 23 (1):3494–3520.

9

Analytical Methods for Cleaning Validation

There are a variety of analytical methods that can be chosen to measure target residues. This chapter will cover analytical methods for chemical residues. The selection of an analytical method for measuring residues is closely tied to the chemical nature of target residues and to the analytical limits established for those residues. Chemical nature includes whether the target residue is organic or inorganic, is soluble in water or other solvents, its degree of polarity, and its stability in the cleaning environment. A key element in the selection of an appropriate analytical method is that the method produces a result that has a logical, scientific link with the target residue [1,2,3]. For example, if the target residue is an organic, nonionized drug active (XYZ), and the acceptance criterion is 2 ppm in the analyzed sample, then using conductivity as an analytical tool would be inappropriate because there is no scientific relationship between the presence of the target residue in the analytical sample and the measurement of conductivity in the test sample. A high performance liquid chromatography (HPLC) method, which was validated to measure XYZ at appropriate levels, would be an acceptable method to choose as an analytical tool for cleaning validation studies.

One can go several steps further, however, and consider conditions under which that HPLC method would be inappropriate for residue testing for validation purposes. For example, if there was evidence that XYZ was degraded during the cleaning process, then that specific HPLC method may not be appropriate for analyzing the target residue. If the HPLC procedure were used for either a swab sample or rinse water sample analysis, the results most likely would be below the detection limit of the method. This would not be helpful information, because if the cleaning or rinsing processes were inadequate, then the species that would be left behind would be the degradation product of XYZ, not XYZ itself.

DETECTION LIMITS

The Food and Drug Administration (FDA) cleaning validation guidelines call for companies to "determine the specificity and sensitivity of the analytical method used" [3]. *Sensitivity* at one time was a useful word for analytical methods (referring to the slope of the working curve); however, in popular usage, it has been loosely used and has become synonymous with either "limit of detection" (LOD) or "limit of quantitation"

(LOQ). The FDA is referring to LOD/LOQ: The LOD/LOQ of the analytical method should be at or (preferably) below the acceptance criterion in the *analyzed sample*. If the target limit in the analytical sample were 5.2 ppm, and a method was only able to detect down to 10 ppm, that method would not be useful for cleaning validation purposes. Because most pharmaceutical manufacturers like to have significant safety built into their processes, they would generally prefer an analytical method with an LOD of at least 25 percent of the target residue limit in the analyzed sample.

The concept of the residue limit in the *analyzed sample* cannot be emphasized enough [4]. As discussed in Chapter 8, the residue limit in the subsequent product is not necessarily the same as the residue limit in the analyzed sample (although the two can be correlated based on batch size, surface area, and sampling procedure). Some companies have established overly restrictive requirements for their analytical methods because they have established requirements for the methods based on limits in the subsequent product rather than in the analyzed sample. In many cases, the residue limits in the analytical sample are considerably higher (by a factor of as much as 10) than the residue limit in the subsequent product. This is due to the "concentration" process that results from the nature of the sampling process. In other words, just because the limit in the subsequent product is 5 ppm, one should not despair because one's analytical method only measures down to 10 ppm. If swabbing is done, for example, the residue limit in the analyzed sample may be on the order of 25 to 50 ppm, and a method with an LOQ of 10 ppm would be suitable without further refinement.

SPECIFICITY

In terms of method *specificity*, there is a natural preference for specific methods. After all, if one has a target residue, the best way to measure that residue is to have an analytical procedure that measures only that species and excludes all potentially interfering species. Specific methods are those methods that target a specific molecule or species and are designed so that possible interferences are eliminated. Specific methods include HPLC, ion chromatography (IC), SDS–PAGE (sodium dodecyl sulfate–polyacrylamide gel electrophoresis), and atomic absorption (AA). With such methods, it is possible to select, for example, column conditions for HPLC such that the target residue is carefully separated from other interfering species. Such methods sometimes involve some kind of chromatographic separation to isolate the target species to be measured.

However, the statement that one should address the specificity of the analytical method used has sometimes been misinterpreted to mean that *only* a specific method can be used. It is unclear where this belief came from, but most likely it came from a misapplication of another FDA position on analytical methods. In the early days of cleaning validation, some companies merely analyzed the rinse water as it exited from a cleaned system. If the rinse water met compendial specifications (such as USP [U.S. Pharmacopeia] Purified Water specifications), those companies considered the cleaning process successful. The FDA objected to this for several reasons [1,3]. One of the concerns was sampling recovery (to be discussed in Chapter 10). Another concern was the fact that the compendial specifications may have no relationship to the presence (or absence) of target residue.

For example, a residue of a potent active may be present in the rinse water in an unacceptable amount, yet the rinse water may still meet compendial specifications. The FDA indicated that it wanted something that could actually measure the target species. An analytical procedure that can specifically measure the target residue is one way of doing this. However, a second way is to use a nonspecific method, so long as the results of that nonspecific measurement can be *directly related* to the target residue.

NONSPECIFIC METHODS

Nonspecific methods are usually methods that measure a gross property that results from contributions from a variety of chemical species. Examples of nonspecific methods include conductivity and total organic carbon (TOC). Each provides a measure of an overall property but provides no information as to the chemical nature of the source of conductance or organic carbon. When a nonspecific method is used for a target residue, it is necessary to make some assumptions about what that nonspecific property represents. This generally involves expressing the property as if *all* the measured property is due to the target species. How is this done? If one is dealing with a target residue that is an organic active, one way is to measure the TOC of the analytical sample. The TOC value is then expressed as if all the carbon present were due to the target organic residue species. If the amount of the target residue calculated by this method is below the acceptance criterion, then it is scientifically sound to say that the residue is less than the acceptance criterion. For example, TOC could be used, and a sample is found to contain

200 ppb carbon. If the target residue were the active in the drug product that contained 25 percent carbon, then that 200 ppb carbon could be expressed as 800 ppb active. An objection could be made that the organic carbon is not, in fact, due exclusively to the target residue, therefore the method is inappropriate. If the objective were to determine the *exact* level of the target residue present, this would be a valid objection. However, the objective is to determine whether the level of the target residue is at *or below* the acceptance level criterion. The organic carbon present is probably not due just to the organic active. There may be contributions from the cleaning agent, excipients (for final dosage forms), or processing aids (for bulk manufacture). However, that is beside the point; these facts only strengthen the case for acceptable residue levels of the organic active. As long as the goal is to determine that the *measured* amount is below the acceptance level, then good science supports using TOC to reach such a conclusion.

Unfortunately, the opposite is not the case; if the TOC measurement indicates that the maximum level of the target residue is *above* its acceptance criterion, then one cannot conclusively say that the target residue is above the acceptance criterion established for that target residue. In such a case, one has to either develop a specific method to confirm the exact amount present or use a more robust cleaning procedure so that the target residue, when measured by TOC with all its related assumptions, is clearly below the acceptance criterion. This, of course, should be worked out in the cycle development work before the actual three process qualification (PQ) runs. If TOC were the only analytical method specified for determining residues, then high TOC values in PQ runs, while not necessarily conclusive evidence of unacceptable residues, would cause the validation protocol to fail.

METHOD VALIDATION

Analytical methods used for measuring residues in cleaning validation protocols should themselves be validated. This validation usually means following standard industry practices for the validation of analytical methods, including evaluation of specificity, linearity, range, precision, accuracy, and LOD/LOQ [5,6,7,8].

Specificity

Specificity is a measure of the validity of the result based on expected interferences. In other words, one needs to confirm whether or not the method can unequivocally measure the target species in the presence of possible interferences. Methods such as HPLC are generally considered specific. However, they are only specific if possible interferences have been evaluated to see if they change the nature of the assay. For cleaning processes, this means that any HPLC procedure should be evaluated to see whether possible residues from the cleaning agent interfere with the assay. Interferences may include changes in retention time, peak height, or peak shape. If cleaning agents are found to interfere in an HPLC assay, the object should be to modify that assay such that the cleaning agent no longer interferes.

Methods such as TOC or an alkalinity titration are generally considered nonspecific because, in most cases, there is more than one species that can contribute to the measured property. Being nonspecific does not mean that the method is unacceptable. What it means is that there is more risk to the manufacturer in meeting their acceptance criteria. The reason is that a nonspecific method must assume a worst case and calculate a target species as if the measured property was all due to that target species. It is a reasonable expectation that at least part of that measured property is due to the interfering species. However, because one cannot specify that percentage, the worst case must be assumed. With a robust cleaning procedure, such an assumption becomes a reasonable risk. It should be noted that the specificity of a method is not an absolute property but is dependent on possible interferences. It may be the case that what is ordinarily considered a nonspecific method, an alkalinity titration, may be a specific method for potassium hydroxide in the cleaning agent if potassium hydroxide is the *only* source of alkalinity in the cleaning process. In fairness to HPLC methods, it should be noted that if interferences are found, the HPLC method may be modified to account for the interference. With assays such as an alkalinity titration, such modifications are generally not possible.

Range

Range is a series of values of the measured species or property over which the analytical procedure was evaluated. It is only necessary to assure that the procedure is valid over a range of expected values. For example, if the calculated acceptance limit for the analytical sample is X ppm, then one might want to evaluate a range from approximately 0.2X to 1.0X. On the other hand, if expected results (perhaps based on pre-qualification studies) are to be in the 0.1X to 0.3X range, then validation of a range of 0.05X to 0.5X may be justified. However, as a practical matter in such circumstances, it makes sense to validate the range up to the 1.0X acceptance limit to cover the possibility that one data point might be obtained in the 0.5X to 1.0X range. Such a scenario is generally not worth the risk of trying to shorten the upper end of the validation range below the acceptance criterion. While it may be interesting to extend the range beyond the acceptance criterion, it is not absolutely necessary. If measured values are obtained larger than 1.0X, the cleaning validation most likely will be unacceptable. Validating the range beyond 1.0X will only confirm to what extent those specific values are unacceptable. Determining of the extent of a valid range for the assay is a matter of risk assessment and will depend on the degree of confidence and expected consistency in any prequalification analytical studies.

LOD/LOQ

LOD is the assay value at which it is still possible to say that the material is present, but it may be not possible to quantify with a specific value. LOD is typically estimated by several techniques. For example, for chromatographic techniques, LOD is estimated at three times the standard deviation of a baseline response. Values that are below the LOD are generally reported as < LOD.

LOQ is the lowest assay value for which a reasonable confidence exists that the value is precise. There are also rules of thumb for estimating LOQs. For chromatographic procedures, the LOQ can be estimated as 10 times the standard deviation of the baseline noise. The LOQ can also be determined experimentally; as a practical matter, it can be considered the lower limit of the validated range of the assay. Any measured value below the LOQ is expressed as < LOQ.

Linearity

Linearity refers to the characteristic of the relationship of the measured property to the level of analyte present. Linearity is an indication that the measured signal is directly proportional to the concentration of the analyte over the range. As a general rule for cleaning validation studies, the expectations are that assays will be linear over the range. Estimates of linearity can be made by such techniques as R^2 determinations (0.99 or better).

Accuracy

Accuracy refers to the trueness of the measurements to known values. This is determined by analyzing known standards. There is no "magic number" for acceptable accuracy. However, more accurate methods are preferred over less accurate methods. For example, if the acceptance criterion was 20 ppm, a method with a accuracy of ± 10 percent, giving a result of 18 ppm, could be considered an acceptable result. On the other hand, a method with an accuracy of ± 20 percent, giving a result of 18 ppm, will be suspect in terms of meeting the acceptance criterion.

Precision

Precision refers to the reproducibility of the method and is often measured by standard deviation. Simple precision is the reproducibility of the results in the same lab over a series of replicate assays using the same operator, the same equipment, and usually on

the same day. Intermediate precision is the reproducibility of results in the same lab using different
operators, different pieces of equipment, and generally done on different days. Ruggedness is interlab reproducibility, involving reproducibility in different
labs. The degree of accuracy required will depend on
the specific situation. If the method is to be developed
in a central lab and then transferred to several remote
locations where analytical support for validation will
occur, ruggedness should be evaluated. For a small
start-up firm, the equipment and analysts may be
limited, and simple reproducibility may be all that is
required. It should be noted that there is inherently
more risk in simple reproducibility, particularly the
risks associated with that analyst leaving the company. It should be noted that in the consideration of
precision, evaluation on more than one instrument,
by more than one operator, or by more than one lab
may not be needed depending on the specific circumstances related to the individual validation protocol. If
the assay is to be used only for validation purposes,
less intensive evaluation is needed. If the assay is to
be used for ongoing monitoring, then a more elaborate evaluation may be needed.

Keys to Method Validation

It should be noted that in many cases, preferences
were given in the discussion of specificity and accuracy. These are not to be considered absolute. In selecting an appropriate analytical method for the
validation task, one must balance a series of needs.
The key is to be aware of the limitations and risks associated with any analytical method and to take steps
to minimize those risks. A robust cleaning procedure

is one way to manage the risks related to analytical methods and residue levels.

It should also be noted that determination of specificity, range, linearity, LOD/LOQ, precision, and accuracy are ordinarily first done on the analytical method itself, independent of the sampling technique. The sampling technique can affect the analytical method. Recovery considerations in sampling techniques using the specified analytical method will be covered in Chapter 10.

TARGET ANALYTES

The analytes targeted for assay will depend on what is targeted in the acceptance criteria. As a general rule, most pharmaceutical manufacturers will have an acceptance criterion for the active ingredient in the equipment cleaned. Therefore, a method to measure that active (either specific or nonspecific) is appropriate. When there is some difficulty in targeting analytes, formulated cleaning agents are often involved. For example, a formulated cleaning agent may contain (in addition to water) a surfactant, an alkalinity source (such as potassium hydroxide), and a chelant. If an acceptance criterion of 10 ppm cleaning agent solids is established, how is that measured? One alternative is to measure each and every species. This makes sense only if an acceptance limit is separately established for each individual component. What is usually done is to target either one component or one property of that cleaning agent formulation [2]. For instance, in the example cited above, it may be possible to analyze for the potassium present, and from that potassium value calculate the total amount of the cleaning agent formulation that might be present. If the cleaning formulation solids contained 45 percent potassium, then a measured level of 0.6 μg potassium would correspond to 1.3 μg of cleaning formulation solids. Such a calculation assumes that the different components of the cleaning formulation are removed from the cleaned equipment at roughly the same rates. While it is possible that there may be differential removal from surfaces, and while it is well known that some surfactants are especially adherent to surfaces, recent work has shown that in a cleaning agent formulation that was freely rinsing, all the components are rinsed at roughly the same proportions within the experimental error of the assay methods used [9]. The concept of "last to rinse" component is a valuable laboratory tool [10]; however, as a practical tool for cleaning validation purposes for determining what component to target in a freely rinsing cleaning formulation, it adds little value.

Alternatively, a gross property such as TOC or alkalinity can be used to measure residues of cleaning agent formulation. Contributed carbon or alkalinity may be due to a combination of components in the cleaning formulation. However, that gross property may be correlated with cleaning formulation solids. For example, a cleaning formulation may contain 9.7 percent TOC on a solids basis. A measurement of 0.30 ppm TOC would correspond to 3.1 ppm of cleaning agent solids. In this particular case, the issue of nonspecificity comes into play. If there are other possible sources of carbon (actives, excipients), then that 3.1 ppm TOC would actually represent an upper limit for the maximum amount of cleaning formulation solids that might be present. In either case, whether a specific component or a gross property of the cleaner formulation is targeted, the assumption is made that what is measured is actually representative of the total formulation.

TYPICAL ANALYTICAL PROCEDURES

Below is a short listing of appropriate analytical procedures and their applicability for cleaning validation purposes. This list is not meant to be exhaustive.

High Performance Liquid Chromatography

HPLC involves injection of the sample into a chromatographic column, separation of the target species from other components in the sample, and then measurement of that target species as it exits the column by ultraviolet (UV) spectroscopy, conductivity, or ELSD (evaporative light-scattering detection). HPLC can generally be tweaked such that it is specific for the target species. The equipment is generally available in pharmaceutical facilities.

Total Organic Carbon

TOC involves oxidation of the sample (by any of a variety of techniques) and measurement of the carbon dioxide generated by either infrared spectrometry or conductance. The method is generally considered nonspecific. TOC usually involves an assumption that all of the measured carbon is due to the target species, and the maximum possible level of the target species is calculated based on this assumption. TOC is becoming more widely used because it is an acceptable technique to replace for the oxidizable substances test for USP Purified Water and because of the possible degradation of actives due to the cleaning environment. For the latter reason, TOC is used commonly in the biotechnology industry for cleaning validation purposes.

Atomic Absorption

Atomic absorption is a specific method for metal ions. It can be utilized in the determination, for example, of sodium and/or potassium that may be present in cleaning formulations. This is not necessarily a common instrument in pharmaceutical analytical laboratories.

Ion Chromatography

Ion chromatography includes specific methods for both anions and cations in cleaning formulations. It can be used to measure both sodium and potassium as cations, and different methods can be used to separate and measure anions, such as the anions from acidic detergents (phosphates, citrates, glycolates) or builders (carbonates, gluconates, silicates, EDTA [ethylenediaminetetraacetic acid]). This is not necessarily a common instrument in pharmaceutical analytical laboratories, but it is becoming more widely used.

Ultraviolet Spectroscopy

For certain surfactants that have a chromophore, UV spectroscopy can be an acceptable tool. The instrumentation is readily available in many pharmaceutical analytical laboratories.

Enzyme-Linked Immunosorbent Assay (ELISA)

ELISA is commonly used in the analysis of protein for the determination of actives. However, because proteins are usually degraded by the harsh conditions (temperature and pH) of the cleaning environment, ELISA has limited practical use for cleaning validation studies.

Titrations

Titrations can vary from alkalinity or acidity titrations, which can be used to give upper level estimates of cleaning agents present, to more specific titration procedures to measure components of cleaning agents, such as titrations for chelants in cleaning agents. The laboratory equipment for these procedures is generally readily available.

Conductance

Conductivity measures a nonspecific property of ions in solution. It can be used as an upper limit estimate of the amount of an alkaline or an acid cleaning agent. Dilute solutions exhibit a linear behavior. If not available, the equipment can be purchased relatively inexpensively.

pH

Some companies have tried to use pH as an estimate of residues of either an alkaline or an acidic cleaning agent. This should generally be discouraged. The measurement of pH in unbuffered systems around neutral is unreliable. In addition, the relationship between the level of cleaning agent and the pH is not a linear one. In such situations, it is preferred to use either conductivity or an acidity/alkalinity titration if a simple analytical procedure is desired for cleaning agent determination. pH can be a useful monitoring tool in that a high or low pH can indicate a system out of control. However, it is not a preferred technique for determining actual levels of alkaline or acidic residues.

REFERENCES

1. FDA. 1998. *Human drug cGMP notes*, vol. 6, no. 4. Rockville, Md., USA:Food and Drug Administration, Center for Drug Evaluation and Research.

2. Gavlick, W. K., L. A. Ohlemeier, H. J. Kaiser. Analytical strategies for cleaning agent residue determination. *Pharmaceutical Technology* 19 (3):136–144.

3. FDA. 1993. *Guide to inspections of validation of cleaning processes*. Rockville, Md., USA: Food and Drug Administration, Office of Regulatory Affairs.

4. LeBlanc, D. A. 1998. Establishing scientifically justified acceptance criteria for cleaning validation of finished drug products. *Pharmaceutical Technology* 22 (10): 136–148.

5. ICH. 1995. Guideline for industry: Text on validation of analytical procedures. *Federal Register* 60:11260.

6. ICH. 1997. Guideline on the validation of analytical procedures: Methodology. *Federal Register* 62:27463.

7. USP. 1995. *United States Pharmacopeia*, 23rd ed. <1225> Validation of compendial methods. Rockville, Md., USA: United States Pharmacopeial Convention, pp. 1982–1984.

8. Kirsch, R. B. 1998. Validation of analytical methods used in pharmaceutical cleaning assessment and validation. In *1998 Analytical Validation in the Pharmaceutical*

Industry, supplement to *Pharmaceutical Technology*, pp. 40–46.

9. Kaiser, H. J., and J. F. Tirey. 1999. Measurement of organic and inorganic residues on surfaces. Paper Presented at Pittcon '99, 7–12 March in Orlando, Fla.

10. Smith, J. 1993. Selecting analytical methods to detect residues from cleaning agents in validated process systems. *Pharmaceutical Technology* 17 (6):88–98.

10

Sampling Methods for Cleaning Validation

As analytical methods are chosen to measure the targeted residue, one must also consider the sampling procedures used as those analytical procedures are applied to cleaned equipment surfaces. The sampling procedure refers to the method of collecting the residues from the surface so that they can be measured and to the selection of which surfaces are targeted for collecting residues for measurements. The objective of appropriate sampling is to end up with analytical results that can be appropriately and logically considered either as representative of the system as a whole or as a worst case in the system (resulting in an upper limit estimate of the maximum residue that could be present). The four types of sampling are direct surface sampling, swab sampling, rinse sampling, and placebo sampling. The features of each will be discussed below.

DIRECT SURFACE SAMPLING

Direct surface sampling involves an analytical instrument directly "applied to" the cleaned surface. The most common example of direct sampling is visual evaluation. In that case, the eye is the means of both analysis and sampling. The key items to consider in visual sampling are listed below:

- *Eyesight of the viewer:* The viewer should have suitable vision, or such vision should be correctable by glasses or contact lenses.

- *The available light for viewing:* Particularly inside equipment, external lighting may have to be introduced to adequately sample the surface. It may also be helpful to use ultraviolet light ("black" lights) to assist in viewing residues that may fluoresce under UV light.

- *The distance of the viewer from the surface:* It is generally the case that the farther away from the viewer, the less sensitive visual examination will be. A surface 1 ft in front of you may appear dirty, while the same surface 12 ft away may appear visually clean.

- *The angle of the light and the viewer to the surface:* These factors can also affect whether a surface is rated clean or dirty.

- *The availability of the surface:* Surfaces that are readily visible can be easily examined. Other surfaces (such as the interior surfaces of ball valves) may require disassembly of the equipment in order to examine them. The availability of other surfaces may be improved by fiber optic scopes, which may allow the visual examination of pipes as long as 7 m.

- *The nature of the residue:* One figure used in the literature for the dividing line between visually clean and visually dirty is 4 μg/cm^2 [1]. This figure can vary depending on the nature of the residue. If it is critical to have a more reliable number, experiments can be performed in which residues are spiked onto model surfaces at different levels (for example, 1, 2, 4, 6, 8, 10, and 12 μg/cm^2). The series is randomized, and a panel of viewers is asked individually to determine which are visually clean and which are visually dirty. The highest level at which all panelists rate the surface as visually clean can then be considered the dividing point. It should be noted that this may vary depending on the nature of the surface (see discussion below). It can also be affected by such factors as the particulate nature of the residue. Particulate residues, such as from powders, can produce visually dirty surfaces at levels considerably below 4 μg/cm^2. For example, a powder can be evenly dispersed across a stainless steel surface at a nominal level of 1 μg/cm^2. If that surface were sampled by swabbing and analyzed, that level would be confirmed. However, that powder would not, as a visual matter, be evenly dispersed across the surface. The residue would exist as "lumps"; in some specific locations, the eye would see residues at levels of 4 μg/cm^2 and above. Therefore, care should be used in relating the visual cleanliness of powder products to specific residue levels. This does not mean that visual examination cannot be used; it still should. However, one should not be surprised if a surface contains a powder residue at a very low level, and yet that surface is unacceptable from a visual cleanliness standard.

- *The nature of the surface:* This includes both the material of construction as well as its surface texture and/or roughness and may be different for different residues. For example, a white powder may be very easy to see on a stainless steel surface, while on a white nylon plastic surface, that same level may be seen as visually clean. Other residues may be readily visible on a polished stainless steel surface, but the same levels of those residues on a rougher surface may be rated as visually clean.

In interpreting results as visually clean, it can be helpful to have a good standard of what visually clean is. For new stainless steel or glass-lined vessels, this may be straightforward. However, if the glass-lined vessel is slightly dulled because of use (or misuse), that dulled surface may be misinterpreted as visually dirty. The same situation may arise involving stainless steel in which the surface is etched (but still clean). Another example that may present problems involves the evaluation of stainless steel welds. In some cases, the welded area assumes a discoloration in the metal itself; such discoloration is not removed by conventional cleaning processes because it is not *on* the surface but rather *in* the surface. In cases like these, it is helpful to have photographs that illustrate a clean surface (in other words, a clean baseline that is the dulled glass surface, the etched stainless surface, or the discolored weld surface) as well as the same surface with sufficient residues to have it rated as visually dirty. These photographs may be used for comparison purposes when evaluating those surfaces for cleanliness.

Care should also be used in interpreting the results of a visual examination of a surface wiped with a wiper or swab. Sometimes, a white wiper is used to accentuate a dark residue, or a black wiper is used to accentuate a white residue. If this method of evaluation is chosen, some thought needs to be put into what the results actually mean. If a surface is visually clean by direct examination with the eye, what does it mean if a fixed surface area is wiped, and the result of the wiping is that residues are visible on the wiping material? Since the wiping process effectively concentrates the residue, this is a distinct possibility. A technique of this type may be useful as a supplement to analytical procedures for cases in which very low residue limits must be obtained. If the wiping procedure is done in a controlled manner, and if what is seen visually on the wiper is correlated with known amounts on the surface (through a series of spiking experiments), such a technique may be used as part of the monitoring process for ongoing cleaning (useful because of its quick turnaround). Use of this wiping technique only for analyzing the cleaned surfaces in the three Process Qualification (PQ) runs for validation purposes probably has limited application. The amount of work that must be done to qualify this method is probably not worth the effort.

In some cases, visual cleanliness alone may be an unacceptable standard. For very low acceptance criteria, a visual examination may tell one that the surface is dirty and therefore unacceptable. However, a visual examination that shows the surface is clean may be inconclusive without specific analytical data. The same applies if there are acceptance criteria regarding microbiological or endotoxin contamination of surfaces; visual cleanliness probably says very little about whether the surface may be acceptable from a microbiological perspective. It is generally expected by the Food and Drug Administration (FDA) that where surfaces can be readily visually examined, they should be examined as part of the PQ process [2].

An alternative method of direct surface sampling is to have a probe that can be placed directly on or over the surface in question. Then the surface is analyzed directly by a technique such as near infrared (NIR) or Fourier-transform infrared spectroscopy (FT-IR) [3]. Such techniques and the associated apparatus to make them useful for pharmaceutical industry cleaning validation purposes are in the developmental stage. It is too early to judge how successful they might be. However, they suffer from one of the same disadvantages as visual examination, namely that the surfaces to which such techniques may be applicable have to be readily available and may be limited in their geometry. More should be available on evaluations of these techniques in the next few years.

SWAB SAMPLING

A swab is a fibrous material that is used to wipe a surface to remove residues from the surface [4,5]. Typically, the swab is a textile fabric of some kind attached to a suitable handle (see Figure 10.1). The swab "head" (the fabric portion) is typically wetted with a solvent (water, an organic solvent, or a mixture), and then is wiped across a fixed surface area of

Figure 10.1. Typical swabs used for swabbing for chemical residues (courtesy of the Texwipe Company LLC).

the surface to be sampled, using a defined wiping motion. The residue is then extracted or desorbed from the swab head into a suitable solvent for subsequent analysis. The key items for the selection of swabbing as a sampling technique are as follows:

- *The nature of the swab:* For microbiological swabbing, the swab head is generally cotton fibers (not a woven or knitted fabric) attached to a handle. In some cases for microbiological sampling, the sampling head is made of calcium alginate fibers. Microbiological swabs are always sterilized prior to use (for obvious reasons).

  Swabs used for chemical residues can vary considerably. However, the most common are made of a knitted polyester fabric head, which is then attached to a plastic handle by a suitable welding process so that adhesives (which could conceivably interfere with any subsequent analytical procedure) are avoided. It is also preferred to use swabs that are "low extractables" grades, here again to minimize interferences. It should be noted that any swab can be used as long it is validated for the residue determination to be made. However, if TOC (total organic carbon) is the analytical technique, then low extractables swabs are required to minimize excessive background carbon due to the swab. While this carbon content from the swab is always accounted for by being subtracted out as background, a low and consistent carbon contribution from low extractables swabs may be critical for consistency and the limit of quantitation (LOQ) for TOC analyses. The swab should also not bind the target residue such that it is not released during the desorption step. This can be confirmed in recovery studies (discussed later in this chapter).

- *Use of solvent for wetting and desorption of the swab:* The solvent used should be appropriate for assisting in the removal of the residue from the surface. In addition, it should be compatible with any subsequent analytical procedure. If the target residue is an active substance and that active is known to be soluble in an organic solvent, then that solvent is a candidate for use in wetting the swab head. Other options include water alone or water adjusted to a high or low pH (the pH depending on whether the solubility of the active increases at either a high or low pH). If TOC is the analytical method, organic solvents should not be used because they will interfere with an accurate measurement of carbon in the residue. The amount of solvent used for desorption may also be critical. A minimum amount of solvent may be necessary for the analytical method. The amount used will also affect the necessary LOQ of the analytical method.

- *The number of swabs used:* Generally, only one or two swabs are used per target surface area. The concept behind using more than one swab is to increase the recovery of the target residue. One swab cannot pick up all of the residues, but it will leave behind a finite amount of solvent, which probably contains some of the target residue. A second swab will pick up an additional amount of the target residue. To enhance recovery, the second swab is sometimes used dry. The theory here is that a dry swab will "mop up" any of the solvent residues left behind, thus increasing recovery as compared to using a second wet swab. While a second swab works well in theory for target residues where the second swab does not interfere with the analytical method, care should be used in adding a second swab to a swabbing procedure where TOC is the analytical technique; the additional background carbon (and the resultant greater variability and larger LOQ) provided by the second swab may negate any benefit due to the additional residue recovered from the second swab, with the net result being little or no overall benefit to the swabbing recovery. Swab recovery studies comparing the use of one swab to two swabs can confirm whether the added swab has any benefit in a given case.

- *The surface area swabbed:* The surface area typically swabbed per site varies from about 25 cm^2 to about 100 cm^2. There is no "magic" number in terms of what is best. The key is consistency, and recovery studies should be done using the same surface area. For flat (or relatively flat) surfaces, there are several techniques to control the surface area swabbed. One is to use a template placed over the surface to be sampled. For example, chemically inert templates made of polytetrafluoroethylene (PTFE), having a "picture frame" of 25–100 cm^2, have been used. The template is held against the surface, and the defined surface is swabbed. An alternative is to train people so that they "know" what the defined surface area is.

- *Swabbed location:* It is also important that the operator know exactly *which* defined area to swab. This is generally identified during scale-up or prequalification trials. The locations selected for swabbing are generally those locations that are the most difficult to clean, representative of different material (e.g., stainless steel, glass, gasket materials), and representative of different functional locations (e.g., side walls, dome, valve, agitator blade, drain). It is most important to identify those locations that are the *most difficult to clean.* If these locations are swabbed, and if the residues in these locations are acceptable, then residues in other locations (easier to clean) should also be acceptable. Also, performing swabbing on representative locations and materials can be helpful in terms of providing a higher degree of assurance in the validation results as well as providing a broader baseline for comparison in case problems should arise in the future. In the validation protocol, the locations for swabbing are either described in writing, marked visually on a schematic of the equipment, and/or attached as photograph of the location (with key features present for orientation), with the sampling area outlined with a marking pen. Such practices make it easier to make sure the proper locations are sampled.

- *The swabbing pattern:* This includes the pattern the swab head makes as it goes across the surface, whether the swabbed pattern is repeated at a 90° angle, as well as any "flipping" of the swab head. (In many cases, the swab head is shaped like a paddle, having two sides; using both sides of the swab is assumed to result in a greater recovery.) One example of a swabbing pattern is shown in Figure 10.2. It is generally a benefit to have both a verbal description of the swab pattern

Figure 10.2. Example of a swabbing pattern

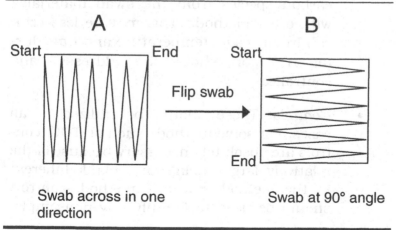

Swab across in one direction

Swab at 90° angle

as well as a visual representation to ensure consistency.

- *Swab handling and transport:* Swab handling includes those steps to prevent contamination during use. Proper handling will depend to a given extent on the analytical procedure utilized. If the analytical method is TOC, then extreme care needs to be used in handling. For example, the swab should be handled with gloved handles, extraneous movements of the swab (waving it through the air) must be minimized, and a careful and clean procedure for cutting the swab head into the vial for desorption must be followed. For an HPLC method, such handling procedures should generally be followed but are not as critical. In addition, transport of the swab in the desorbing solvent from the time and site of swabbing to the time and site of performing the analytical assay must be considered. Here again, with some assays such as TOC, this may need to be carefully controlled (because of the extraction of carbon species from the swab materials); with other methods, this may be less critical. In any case, temperature and time during transport should be addressed and controlled.

- *Controls:* These should be utilized in all cases. For some methods such as TOC, controls are absolutely necessary because of the relatively large background blank inherent in the method. For any method, controls should be used to identify any nonsamplerelated sources of error. The sampling control should be prepared using the same swab, solvent for desorption, and type of vial. The sampling control should be prepared at the same time as the experimental sample is taken. The sampling control should not be prepared in the laboratory; as much as possible, the sampling control should have the same history as the experimental sample swab, except that it does not touch the equipment surface. It should be noted that for any analytical procedure, there may be other controls prepared in the laboratory; their use may be necessary but should be in addition to the sampling control.

All of these factors—the swab itself, the number of swabs, the type of solvent used for wetting, whether they are all wet or a combination of wet and dry, the surface area to be swabbed, the swabbed location, the swabbing pattern, swab handling and transport, and swabbing controls—should be identified in the swabbing SOP (Standard Operating Procedure) and/or validation protocol. The SOP is not only the procedure to follow in the validation protocol; it is also followed exactly when the swab recovery studies are done.

RINSE SAMPLING

Rinse sampling involves using a liquid to cover the surfaces to be sampled. There are really two cases of this: single point final rinse sampling and a sampling rinse separate and distinct from the final rinse [6]. Before each is discussed, it should be pointed out that rinse sampling has a negative connotation for validation purposes. Early in the history of cleaning validation, some companies misused rinse sampling. One misuse involved trying to assert that as long as USP (U.S. Pharmacopeia) Purified Water was used for the rinse, if the final rinse water met the USP specifications (or if the rinse water was unchanged from the original specifications), then the cleanliness of the equipment was validated. The FDA objected to this (as rightly they should have) because there was no measure of any target residue. It should be clear, particularly with potent drugs, that residues of those drugs may be present in the rinse water at unacceptable levels, yet the water could still meet the USP specifications. A second objection the FDA had to rinse sampling *as it was done at that time* was that manufacturers were measuring a target residue in the rinse water, but they had not validated that those residues could have been detected in the rinse water had they been present on the equipment surfaces at the time of rinsing. This is the so-called "dirty pot" analogy; if one wants to determine the cleanliness of the pot, one examines the pot, not the wash or rinse water [2]. This is a good analogy, but it does have its limits. It is possible to examine the rinse water if it can be demonstrated, through recovery studies, that any residues "on the pot" are readily removed by the rinsing process. There are two keys to overcoming objections to rinse sampling as it was done: (1) make sure that what is measured in the rinse sample is directly related to the target residue, and (2) make sure that the rinse procedure has been demonstrated to remove the target residue from model surfaces in valid recovery studies. With that in mind, the following is a discussion of the two types of rinse sampling procedures.

Single Point Final Rinse Sampling

In single point final rinse sampling, a sample (usually a "grab" sample) is obtained at the end of the final rinsing process. If the rinsing medium is water, then it is a water sample; if a solvent is used for the final rinse, then it is a solvent sample. The volume of rinse sample taken will depend on the analyses done but typically is in the 500–1,000 mL range. The sampling point is usually identified as that point at the end of the cleaning circuit at which the water exits the cleaning circuit. This is the most common method of rinse sampling. Final rinse sampling, providing the two conditions discussed above are addressed, has the advantage of being a simple procedure. It also can be an effective means of sampling sites that are not visible or are not readily sampled by swabbing, including process pipes. Rinse sampling of this type also provides an *overall* measure of the contamination of a system; if it is reasonable that the contamination is uniformly dispersed throughout the subsequently manufactured product, then rinse sampling may give a valid measure of the overall potential contamination of the subsequent product.

A key question is relating what is measured in the rinse water to possible contamination levels of that residue in the subsequent product. If certain (reasonable) assumptions are made about what a final rinse sample represents, then it is possible to assert that, in the worst case, that a level of, for example, X ppm in the final rinse correlates to no more than (that is, an upper limit of) X ppm in the subsequent product. If this approach is taken, it is useful to clarify those assumptions in a justification document.

Separate Sampling Rinse

A separate and distinct rinse is performed with a fixed amount of sampling medium after the process rinsing (as part of the cleaning SOP) is completed. The only function of the sampling rinse is to sample the equipment for analytical purposes. This rinse sampling procedure can either be done on a batch or continuous basis. The only additional criterion for using it for residue calculation sampling procedures is that the rinse procedure contact (and therefore sample) *all* surfaces of the equipment. This is a requirement because in rinse sampling, it is not possible to focus the rinse solution on the most difficult-to-clean surfaces (as might be done with swab sampling). Because of this limitation, all surfaces must be sampled to obtain a snapshot of the overall contamination.

This separate rinse sampling has several advantages. One is that it is possible to use a sampling rinse solution that is different from the process rinse. In aqueous processing, the final process rinse is usually just water. If the active agent has poor solubility in water but is more readily soluble at a high pH, for example, it may be possible to use a dilute solution of sodium hydroxide as the sampling rinse to enhance and assure the recovery of that active agent residue. Another alternative would be to use an isopropanol/water mixture as the sampling rinse solution.

A second advantage is that because the sampling rinse has a fixed volume, it is possible to more carefully approximate the possible contamination of the subsequent product. As a practical matter for discrete rinses (dump and fill systems), in which the amount of sampling rinse solution is approximately the same as the amount of the next product, a level of X ppm of target residue in the sampling rinse correlates to

approximately X ppm of the target residue in the subsequent product (assuming uniform distribution). For
continuous rinsing processes, as in clean-in-place
(CIP), it is likely that the volume of the sampling rinse
to sample all parts of the equipment is considerably
less (as much as 80–90 percent less) than the volume
of the subsequently manufactured product. If this is
the case, a level of 1.0X ppm of the target residue in
the sampling rinse correlates with approximately
0.1X to 0.2X ppm of the target residue in the subsequent product. In this continuous CIP sampling rinse
procedure, it is possible to leverage the analytical procedure (much as is done in swab sampling) such that
relatively high numbers in the analyzed samples actually represent significantly lower levels of potential
contamination in the subsequent product.

PLACEBO SAMPLING

Placebo sampling involves manufacturing a placebo
batch (the drug product less the active drug substance) of the subsequently manufactured product in
the cleaned equipment. Following manufacture of the
placebo product, the placebo is analyzed for the target residue, for example, the active in the drug product that was cleaned from the equipment. If that
active is found in the placebo at a certain level, say
X ppm, then it is assumed that the contamination of
that subsequently manufactured product with the active would also be at a level of X ppm. If this level is
below the acceptance criterion in the subsequent
product, then the equipment is assumed to be adequately cleaned.

The FDA has expressed objections to placebo
sampling [7]. One objection is that it assumes that the

contaminating residue will be uniformly dispersed throughout the placebo batch; so when a sample of the placebo is taken and analyzed, that sample will be representative. This is a valid objection, but uniform contamination is an assumption behind swab and rinse sampling also. A second FDA concern is the analytical power to adequately measure target residues in the placebo matrix. It is difficult enough to measure an active at very low levels in pure water or pure solvent. Measuring that active in the presence of the components of the placebo may present an analytical challenge. One additional item to be considered is a valid recovery study to demonstrate that the placebo procedure could actually recover the target residue if it were on a surface. The FDA's position in their guidance document is that placebo sampling is acceptable provided it is supplemented by either swab or rinse sampling. For this reason, placebo sampling is seldom done.

However, one area in which placebo sampling may have value is in the evaluation of nonuniform contamination. This does not refer to nonuniform contamination of the equipment surfaces because in most cases equipment surfaces are not uniformly contaminated. Rather, it refers to nonuniform contamination of the subsequently manufactured product. There may be situations in which the contaminating residues on the cleaned equipment are preferentially removed by and thus contaminate only a portion of the subsequently manufactured product. For example, in continuous filling equipment, it is possible that the contaminating residues on the filling equipment surfaces would be removed by the first portion of product to be filled. This would result in the first vials filled having a higher level of contamination, with subsequent vials having lower levels of contamination. In such a situation, it may be possible that the vials initially filled would have unacceptable levels of contaminants, while subsequent vials would have acceptable levels of the same residues. Placebo sampling can address this by filling with a placebo product using the cleaned equipment and analyzing vials on a regular basis (for example, vials 1, 10, 50, 200, 1,000) for the target residue. If vial 10 had an unacceptable residue level but vial 50 had an acceptable level, intermediate vials could be analyzed to determine with which vial the residue became unacceptable. A safety factor could be added to this, and a number of initially produced vials destroyed. An alternative to this is to select a more rugged cleaning process, such that any nonuniform contamination does not produce unacceptable contamination. If the non-uniform contamination of the subsequent product cannot be assumed to be predictable (such as the random contamination of vials), then placebo testing offers little help (other than helping to identify the problem). In that case, a more rugged cleaning process is called for.

Another disadvantage of placebo testing is that it does not allow leveraging of the analytical method, as does swab testing and certain types of rinse sampling; what is measured in the placebo product is assumed to be the contamination level of the subsequently manufactured product. In contrast, with a swab sample, it may be possible that a relatively high value of an analyte (e.g., 25 ppm) in the analyzed sample may correlate with a much lower value (e.g., 4 ppm) of possible contamination of the subsequent product.

RECOVERY STUDIES

A key in any sampling procedure is that the sampling procedure quantitatively remove the residue from the surface for subsequent analysis [8,9]. This is demonstrated in so-called recovery studies. In a recovery study, a fixed amount of the target residue is spiked onto a model surface, the surface is sampled with the sampling procedure, and the sample obtained is analyzed by the analytical method. The amount of target residue is then expressed as a percentage of the amount spiked to give the "percent recovery." This percentage must then be used to adjust the analytical results obtained to accurately reflect potential contamination. For example, if the contamination of a surface in a validation study is measured at 12 μg per swab, and the recovery with that swabbing procedure and that residue is 75 percent, then the potential contamination is 12 μg ÷ 0.75, or 16 μg per swab.

Sampling recovery procedures are usually done in the laboratory. Some of the keys to successful swab recovery studies are as follows:

- *The model surface used:* This should reflect actual conditions. For example, if stainless steel is the swabbed surface in the process equipment, the contaminating residue should be spiked onto stainless steel coupons. Such coupons should be of sufficient size such that the area swabbed is the same as the area specified in the swabbing SOP.

- *The amount of the contaminating residue:* The amount of residue, in μg/cm^2, should be in the range of that expected to be found in actual validation trials. This level should definitely be below the acceptance criterion specified for contamination. If recovery is expected to be different depending on the level of contamination, then a series of recovery studies should be done. For example, recovery studies could be done at contaminating levels equal to about 10 percent, 30 percent, and 60 percent of the acceptance criterion, or they could be done at 1×, 2×, and 4× the LOQ of the analytical method. Either the lowest percent recovery obtained could then be safely used for adjusting the actual analytical results, or a series of recovery percentages could be used depending on the analytical result obtained.

- *The nature of the residue:* Is the spiked residue sampled as applied, after drying, or after baking at a specified temperature? The nature of the residue should approximate that in the equipment to be cleaned. If the equipment is steamed prior to cleaning, then it is appropriate that the spiked residue be treated in a similar manner so that the recovery study approximates as much as practical the actual conditions in the validation swabbing procedure.

- *The swabbing SOP:* The SOP should be exactly the same SOP used in the validation protocol swabbing. This includes all of the issues discussed previously regarding swabbing.

Rinse sampling recovery studies can be done in the laboratory; however, it is very difficult to simulate rinse sampling conditions, particularly a CIP rinse, in a laboratory study. The approach for rinse sampling recovery studies is to design a reasonable laboratory study using worst-case conditions. It is then assumed that recovery under actual sampling rinse conditions would be no less than, and most likely greater than, the percentage recovery obtained in the worst-case lab study. Rinse sampling studies can be simulated by spiking the target residue on the bottom of a stainless steel beaker and then applying the sampling rinse in a manner to cover the surface with a fixed agitation (usually minimal) and a fixed time (usually less than the expected contact time expected in the actual rinsing on the process equipment). An alternative is to spike a coupon with the target residue and do a simulated rinse by allowing a fixed amount of sampling solution to flow over the coupon into a collection vessel for analysis.

Issues addressed in swab sampling, such as the nature of the surface and the amount and nature of the target residue, also apply to rinse sampling recovery. One additional critical parameter is the relative ratio of the sampling solution to the surface area sampled. This should approximate that found in real-life processing; as a worst-case, the ratio of the amount of sampling solution to the surface area sampled should be less than the real life situation.

Acceptable percentage recoveries are > 50 percent*. Recoveries above 80 percent are preferred, but it is recognized that recoveries at the very low levels

---

*Answer to audience question given by Y. Henry of FDA, at AAPS Workshop on Current Issues: Analytical Validation for the Pharmaceutical Industry, 6–7 April, 1998. In Arlington, Va.

present in cleaning validation situations can add considerably to the low recovery percentages.

It should also be noted that certain sampling procedures may be considered invasive, such that special cleaning or a repeat of the cleaning process may be required before the equipment can be used again for manufacturing products.

REFERENCES

1. Fourman, G. L., and M. V. Mullen. 1993. Determining cleaning validation acceptance limits for pharmaceutical manufacturing operations. *Pharmaceutical Technology* 17 (4):54–60.

2. FDA. 1993. *Guide to inspections of validation of cleaning processes*. Rockville, Md., USA: Food and Drug Administration, Office of Regulatory Affairs.

3. Biwald, C. E., and W. K. Gavlick. 1997. Use of total organic carbon analysis and Fourier-transform infrared spectroscopy to determine residues of cleaning agents on surfaces. *Journal of AOAC International* 80 (5):1078–1083.

4. Lombardo, S., P. Inampudi, A. Scotton, G. Ruezinski, R. Rupp, and S. Nigam. 1995. Development of surface swabbing procedures for a cleaning validation program in a biopharmaceutical manufacturing facility. *Biotechnology and Bioengineering* 48:513–519.

5. Cooper, D. W. 1997. Using swabs for cleaning validation: A review. In *cleaning validation: An exclusive publication*, edited by B. Anglin. Royal Palm Beach, Fla., USA: Institute for Validation Technology, pp. 74–89.

6. LeBlanc, D. A. 1998. Rinse sampling for cleaning validation studies. *Pharmaceutical Technology* 22 (5):66–74.

7. FDA. 1998. *Human drug cGMP notes*, vol. 6, no. 4. Rockville, Md., USA: Food and Drug Administration, Center for Drug Evaluation and Research.

8. Kirsch, R. B. 1998. Validation of analytical methods used in pharmaceutical cleaning assessment and validation. In *1998 Analytical Validation in the Pharmaceutical Industry,* supplement to *Pharmaceutical Technology,* pp. 40–46.

9. Chudzik, G. M. 1998. General guide to recovery studies using swab sampling methods for cleaning validation. *Journal of Validation Technology* 5 (1):77–81.

11

Microbial Issues in Cleaning Validation

A rigorous evaluation of microbial contamination for cleaning validation has had a low profile during the first decade of cleaning validation [1]. The focus, at least in terms of acceptance criteria, has been on chemical residues, specifically residues of the active and the cleaning agent. The Food and Drug Administration (FDA) guidance document on cleaning validation specifically says that the guide is "intended to cover cleaning for chemical residues only." Despite that main concern with chemical residues, the guidance document also states that "microbial aspects of equipment cleaning should be considered." The emphasis in the guidance document, however, relates to microbial proliferation during storage [2].

IMPORTANCE OF MICROBIAL RESIDUE CONTROL

The rationale for considering microbiological contamination as part of a cleaning program is that microbiological quality is of concern for all products [3]. At a minimum, the absence of certain enteric organisms in the cleaned equipment is a must for all products. For biotechnology manufacturing involving microbial cells, the presence of species other than the one desired for manufacturing purposes may interfere with productivity and quality. For certain products, such as those that are subsequently sterilized, the issue of bioburden may be critical to the validation of those sterilization processes. The issue of microbiological contamination of cleaned equipment for sterile products may also be related to endotoxin levels in the subsequently manufactured product. Additional concerns are the possible effects of microbes on the stability of the drug product or perhaps on the bioavailability of the drug active.

Thus, while the FDA guidance document does not emphasize microbial residues as part of the acceptance criteria for cleaning validation studies, many companies will include microbiological considerations as part of the acceptance criteria in cleaning validation protocols.

CHANGES IN MICROBIAL RESIDUES

For drug product manufacture, one key way in which microbial residues are different from chemical residues is that chemical residues generally are transferred unchanged to the subsequently manufactured product. If one has a certain amount of chemical residue X on the cleaned equipment, one can readily calculate the potential contamination of the subsequently manufactured product (assuming uniform distribution). With microbial contamination, the situation is slightly different. As pointed out in the FDA guidance document, there may be microbial proliferation after the cleaning step (and during storage), which may considerably change the level of microbial residues. On the other hand, there may be microbial death on cleaned, dried equipment that significantly lowers the bioburden transferred to the subsequently manufactured product. The main issue is that levels of microbial residues present at the end of cleaning are not as relevant as levels of microbial residues that are present at the beginning of manufacture of the next product (although the two are definitely related). The key point is that microbial residues may change during storage of the cleaned equipment. This is why the FDA guidance document emphasizes the need to focus on equipment drying before storage.

The level of microbial residue can change during storage, and the level present at the beginning of manufacturing is not necessarily predictive of what levels may be in the subsequently manufactured product. The nature of that subsequently manufactured product (as well as storage of that product) may change the level of microorganisms present. If the product is a dry product with a water content of < 0.6 percent, it is generally accepted that microbes will not grow in such products [4]. On the other hand, if the product is a neutral aqueous product (with no preservatives), one can expect microbial proliferation. If that same product had a preservative or was formulated with a significant amount of alcohol, one could expect either a minor change in microbial content or even a total reduction in microbial count over a short period of time (e.g., one week). Clearly, the situation is more complex than with chemical residue contamination. This necessitates a more careful review of specific situations for the proper consideration of microbial residue limits in cleaning validation studies.

MICROBIAL RESIDUE REDUCTION

The good news is that the control of microbial residues in properly engineered manufacturing equipment is a fairly easy process. Effective cleaning by itself can in most cases produce equipment that will have acceptable microbial residue levels. Effective cleaning in this case generally means cleaning with a hot solution of a highly alkaline or acidic cleaning agent containing surfactants, which also remove the chemical residues that are present in the system. The purpose of the acidity or alkalinity is to produce a cleaning environment that is hostile to microorganisms. The same is true for using an elevated temperature (generally above 60°C) to produce an environment hostile to microbes. The purpose of the surfactant is to assist in wetting, for better contact with the microbes, as well as to assist in the "carrying away" of the microbial residues. It is important that the cleaning process also remove the chemical residues present; significant amounts of chemical residues left behind may serve as microbial traps, which prevents the destruction or removal of the microorganism during the cleaning process.

The bad news is that proper design and engineering of the equipment is necessary to eliminate locations where microorganisms can hide and proliferate. These are generally the same locations that may cause problems with chemical residues, including dead legs and crevices (such as at the junction of a stainless steel part with a gasket). In addition, as mentioned earlier, anywhere that water can pool is a potential source of microbial proliferation.

Expectations are not that the equipment should be sterile after the cleaning step, particularly if the final rinse water was USP (U.S. Pharmacopeia) Purified Water or Water for Injection. Any microbes left from the final rinse would still be in the system, unless they die following drying of the equipment. Using a chemical disinfectant or sanitizer (such as peracetic acid or a quaternary ammonium chloride) may provide a further reduction in microbes, but the use of these products would require a rinse, thus possibly reintroducing microbes from the final rinse water. Other chemical disinfectants or sanitizers, such as hydrogen peroxide or alcohol, may be used for further microbial reduction, since they may not require a rinse (providing adequate time is allowed for the hydrogen peroxide to decompose or for the alcohol to evaporate before the manufacture of the next product). The use of steam in a SIP (sterilization-in-place or steam-in-place) procedure may provide a further reduction or even produce sterile equipment if properly validated [4]. The need for using a microbial reduction agent after the cleaning step should be evaluated on a case-by-case basis. If acceptable microbial quality can be achieved by cleaning alone, the use of an agent for further microbial reduction should generally be avoided.

In terms of establishing specific acceptance levels for microbial contamination, the levels established will depend to some extent on whether the product is sterile (subsequently sterilized), aseptically produced, or nonsterile. Each of these cases will be considered separately.

Limits for Subsequently Sterilized Products

The key for subsequently sterilized products is that the maximum bioburden present in the sterilized product must be specified for sterilization validation. Therefore, the contribution of bioburden from the equipment itself must be such that it is consistent with the bioburden specified for sterilization validation. Unfortunately, the cleaned equipment is not the only source of bioburden, so other sources of bioburden (such as raw materials and packaging components) must also be addressed. The principle behind setting limits for acceptance testing for validation purposes is that the total of the bioburden from all sources should be less than or equal to the maximum specified in the subsequent sterilization validation. There are several issues here. One is the chicken-and-egg question: Which is specified first, the sterilization validation bioburden or the microbial limits following cleaning? The answer, of course, is that they are selected at the same time, since both have to be reasonable and achievable. Good science and a knowledge of the system through prequalification testing is usually an acceptable guide. The second question is, What is the distribution of that maximum bioburden among the various contributing sources? The bioburden can usually be allocated based on data that are available from prequalification testing (cleaning residues, raw material bioburden, and packaging bioburden). Any worst case, with all contributors at their maximum, should be at or below the maximum bioburden specified for sterilization. Because of inherent variability in microbiological testing, some safety factor should be used to insure that in any worst case, the total bioburden is well below that maximum level.

In calculating the contribution of the microbial residues in the cleaned equipment to possible bioburden of the subsequently manufactured product, a calculation similar to that used for chemical residues is used to calculate the colony forming units (CFU) per volume or weight of the subsequently manufactured product:

$$bioburden = \frac{A \times B}{C}$$

where A = CFU per surface area, B = product contact surface area, and C = volume or weight of product.

This calculation can be used assuming uniform contamination of the subsequently manufactured product. In addition, any change in the bioburden once the microbes are in the subsequently manufactured product must be considered.

Limits for Aseptically Produced Products

The limits for aseptically produced products is not unlike that for subsequently sterilized products, except it is the *equipment* (rather than the product) that is subsequently sterilized (or decontaminated). The main issue here is that validation of the equipment sterilization process must take into account the bioburden present in the equipment. The microbial bioburden acceptance criteria following cleaning must be consistent with the bioburden used for sterilization validation purposes. A second issue is the possible effect of that bioburden on endotoxin levels. If the equipment is steam sterilized, there is still a concern about endotoxins remaining in the equipment. This is probably not a significant issue for most cases in which thorough cleaning is done. However, it should not be overlooked.

Limits for Nonsterile Products

The issue for nonsterile products is relatively straightforward. Manufacturers of nonsterile products realize that there will be some microbial load in their products. However, microbial loads consistent with the microbial content (species and number) of other products that a consumer might utilize (for example, drinking water) may be reasonable standards. At this time, the USP has a proposed revision of <1111> "Microbial Content of Nonsterile Pharmaceuticals," which can provide guidelines for what is acceptable in the finished product [5]. The equation presented earlier for determining the contribution of microbial residues in the cleaned equipment to the bioburden in the finished product can be used. As in that situation, other contributors (raw material, packaging, environment) to bioburden of the finished product must be considered in establishing the acceptance criteria for the cleaned equipment.

WHAT IS MEASURED FOR ACCEPTANCE CRITERIA?

The acceptance criteria for microbial residues in a cleaning validation study, for any of the three cases covered above, are relatively straightforward compared to determining the acceptance criteria for chemical residues. The acceptance criteria preferably include a specification of restricted organisms (that is, those that should not be present at all, such as enteric organisms) and a specification of the maximum number of CFUs per swab or contact plate. In addition, during actual testing, an identification of the species present should be made.

The maximum number acceptable in the cleaned equipment should be based, as discussed above, on the contribution of that microbial load to the subsequently manufactured product or to the bioburden for the subsequent sterilization process.

In selecting acceptable limits for cleaning validation acceptance criteria, one possibly could view these numbers as extremely high; in most cases, one would never have equipment contaminated at those levels. However, it is necessary to distinguish what is a scientifically justifiable acceptance limit and the action/alert limits for routine process monitoring. In almost all cases, the action/alert limits measured as part of routine monitoring will be *significantly* below the acceptance criteria for cleaning validation purposes. This is to be expected. The acceptance criteria for cleaning validation purposes are based on a scientifically (and logically) sound basis for what could contaminate the system and still be acceptable. The action/alert limits are initially established as tentative limits based on prequalification testing. As more data are obtained from routine monitoring, these action/alert limits may be revised. The purposes of action/alert limits for routine monitoring are different from the acceptance criteria for cleaning validation purposes. The action/alert levels for monitoring purposes are designed to provide an early warning of possible changes in the system through a review of trend data. They are not necessarily indicative of a quality problem with the finished product. This contrast between validation acceptance criteria and routine monitoring limits is important for chemical monitoring as well as for microbial monitoring.

SAMPLING TOOLS

The methods for microbiological analysis are well established [6]. Sampling is done with either sterile swabs (cotton or alginate tipped) or contact plates or by sampling the rinse water in a membrane filtration procedure. Both swabs and contact plates require access to the area being sampled. Contact plates generally can access surfaces that are relatively flat. Unfortunately, those areas most likely to be contaminated with microorganisms are not flat surfaces but rather more complex surfaces such as drains, fittings, and gaskets. For this reason, swabs are generally necessary for validation purposes. With any surface sampling technique, consideration must be given to the adequate removal of any sampling media (e.g., agar, alginate) introduced onto the sampled surface during sampling. Testing of the rinse water by membrane filtration techniques is relatively straightforward. However, rinse water testing depends on the assumption that microbes attached to surfaces can readily be removed by a flowing stream of water, a questionable assumption at best. One researcher has estimated the recoveries of the different sampling methods as ~50 percent for contact plates, 25–50 percent for swabbing, and <25 percent for rinse sampling [7].

OTHER CONCERNS

It should also be noted that microbial testing for cleaning validation usually focuses on bacterial or mold/yeast contamination. While viral contamination and contamination from prions (admittedly, neither truly qualifies as microorganisms) are of concern, methods to assay them are problematic. Traditionally, prions and viruses have not been included in any microbial monitoring schemes.

REFERENCES

1. Docherty, S. E. 1999. Establishing microbial cleaning limits for non-sterile manufacturing equipment. *Pharmaceutical Engineering* 19 (3):36–40.

2. FDA. 1993. *Guide to inspections of validation of cleaning processes.* Rockville, Md., USA: Food and Drug Administration, Office of Regulatory Affairs.

3. FDA. 1998. *Human drug cGMP notes,* vol. 6, no. 1. Rockville, Md., USA: Food and Drug Administration, Center for Drug Evaluation and Research.

4. Baseman, H. J. 1992. SIP/CIP validation. *Pharmaceutical Engineering* 12 (2):37–46.

5. USP. 1999. United States Pharmacopeia, <1111>, Microbiological attributes of nonsterile pharmaceutical products. *Pharmacopeial Forum* 25 (2):7785–7792.

6. Clontz, L. 1998. Microbial limit and bioburden tests: Validated approaches and global requirements. Buffalo Grove, Ill., USA: Interpharm Press, pp.129–144.

7. Docherty, S. E. 1999. Microbial cleaning issues. Paper presented at the First International Conference on Cleaning Validation, 12–13 April in Princeton, N.J.

12

Change Control and Revalidation

Once the cleaning process is validated, it should be operated under change control procedures, and the validation should be confirmed on a regular basis [1,2]. The cleaning validation master plan should specify that validated cleaning procedures are operated under change control. There should be a change control SOP (Standard Operating Procedure) that specifies the responsibilities, procedures, and templates for documenting information related to changes. The cleaning validation master plan should also specify the frequency of "revalidation" and the requirements for revalidation.

CHANGE CONTROL

Change control procedures for cleaning validation are no different in principle than change control for any other validated process. The purpose of change control is to document any changes in the cleaning process and to evaluate the effects (if any) of those changes on the cleaning process. Changes to the cleaning process could be unintentional (e.g., the failure of a recirculation pump in a CIP [clean-in-place] system), or intentional (e.g., changing the rinsing time of the cleaning procedure).

It is clear that all changes should be documented. What is critical is the level of justification required to specify that the change has no effect on the validated process. In all cases, the judgment of a professional qualified to make that scientific assessment is required. This judgment usually involves the evaluation of some data to determine whether or not the change is significant. That data may involve a simple Installation/Operational Qualification (IQ/OQ) in the case of like-for-like equipment changes (e.g., a new pump in the CIP circuit). It may also involve the development of prequalification experiments and/or Process Qualification (PQ) runs to confirm performance (e.g., in the case of a change in the rinsing time).

As with other aspects of validation, there is an element of risk management in terms of what level of data is sufficient to support a declaration that the process is still validated. Most scientists could probably agree that, at a minimum, a certain amount of work would be required (but would not necessarily be sufficient) to support a change. At the other extreme, most scientists would clearly agree that a certain (rather exhaustive) list of data would clearly be sufficient to justify a change. Between the two is a middle area where scientists in different manufacturing situations and corporate environments could make differing decisions.

Below is a discussion of some situations and issues that should be considered in selected cases of changes involving a validated cleaning process. This list is not designed to be exhaustive nor is it designed to lay down hard and fast rules for what data need to be gathered. It is designed to bring up points that should be considered in evaluating specific changes. The key to any of these changes is to document the change and to document the evaluation considerations utilized in deciding whether the change is significant.

Change in Pumps, Spray Balls, and/or Other Mechanical Equipment

In like-for-like changes, the same model and manufacturer of a pump or spray ball, then the change can be made at a minimum with a simple IQ procedure. While regulatory requirements [3] are such that OQ is not required, some manufacturers may decide to perform OQ on like-for-like equipment changes as part of their risk management. If the changes involve equivalent equipment from a different manufacturer, then IQ and OQ should be done. For the pump, OQ may involve measuring the flow rate (in gallons per minute). In a spray ball change, OQ would involve measuring the pressure at the spray head and spray distribution by riboflavin testing. The nature of the differences would come into play as one decided whether some elements of PQ runs should be done before accepting that the change does not affect the validity of the initial validation. For example, an equivalent spray ball with acceptable coverage may or may not give equivalent cleaning, depending on the specific locations of the holes in the spray ball.

Change in Water Quality
in the Washing Step

As a general rule, it is fairly easy to justify that a change in water quality from a less pure water quality to a more pure water quality is not a significant change as far as cleaning validation is concerned. If a cleaning process is validated with USP Purified Water and the system is upgraded to WFI (Water for Injection), the effect on the cleaning process should be inconsequential. Therefore, provided the WFI system is validated and the change documented, no further data are required. It is sufficient that a professional judgment is made that the cleaning process is unaffected. The same would apply to a cleaning process that changed from a potable water system to a deionized or Purified Water system. The only caution is that under certain conditions, the presence of hard water ions may serve to defoam the system. For example, a CIP system that worked fine with tap water could experience problems with foaming, and hence pump cavitation, with deionized water. This may require a simple laboratory evaluation of the effects of different water qualities on the foaming properties of the cleaning agent.

On the other hand, a change from a more pure water quality to a less pure water quality in the washing step may involve the development of some data, and perhaps even one PQ run, to confirm that no change in the cleaning process has occurred. If the change is from a water quality containing "no" hardness ions (such as deionized, Purified, or WFI water) to tap water, a more elaborate evaluation is required. One factor to consider is the effect of hardness ions on the surfactants in any cleaning agent used. It is well documented that the effectiveness of most surfactants is decreased by the presence of hardness ions such as calcium and magnesium. A system that cleans well with deionized water may not clean as well when tap water is used in the washing step. This evaluation may involve some laboratory studies and at least one PQ run to determine equivalent cleaning.

Another possible effect of switching to tap water in the washing step is the effect of alkalinity on the calcium ions, involving the formation and precipitation of calcium carbonate. Unless controlled by chelants in the cleaning solutions or by a subsequent acid wash, these calcium carbonate precipitates may appear as white deposits on equipment surfaces.

Change in Water Quality in the Rinsing Step

As with the washing step, a change from a less pure water quality to a more pure water quality in the rinsing step can usually be made without further evaluation, provided that the more pure water system is validated as a water system. As with the washing step, the change from a more pure to a less pure quality rinse water requires more substantial evaluation. For example, the change from WFI to Purified Water may require an evaluation of the microbiological quality of the surfaces after the cleaning process. One should also evaluate such a proposed change in light of the general rule mentioned in Chapter 5: The quality of water used for the final rinse should be at least as good as the quality of water added to the vessel for the subsequent manufacturing step.

Change in the Cleaning Agent

A change in the cleaning agent can be of two types. One is a minor change in the components of a formulated cleaning agent. A minor change may be a change in the amount of one component or perhaps the change of one component. Examples of component changes may be a change from sodium hydroxide to an equivalent (on a molar basis) of potassium hydroxide or a change from one acrylate polymeric dispersant to another of slightly different molecular weight or structure. Such a change will generally involve some laboratory studies to demonstrate equivalent performance and may include at least one repeat of a PQ run to determine that there is no significant effect from the minor change. Such a change should also cause one to investigate both the supplier's rationale for such a change (including an investigation of that supplier's change control and notification procedures) and the reliability of the supplier in supplying products for validated cleaning applications.

A second case involves a complete change in the cleaning agent. This could involve a change from one cleaning solvent to another, from a commodity chemical alone to a formulated cleaning agent (in aqueous processes), or from one formulated cleaning agent to another formulated cleaning agent with different components and/or ratios of components. These changes ordinarily would be considered new cleaning processes and would require revalidation, including three PQ runs with the new process.

Change in the Manufacturing Process of the Cleaned Product

A change in the manufacturing process for the drug product cleaned may cause a change in the nature of the residues to be cleaned or a change in the worst-case cleaning locations. For example, an increase in the processing temperature may result in residues that are baked on surfaces or are dried on surfaces in larger quantities than with lower temperature processing. Such changes in processing may thus result in more difficult-to-clean residues. A processing change that results in a significant change in the nature or location of residues to be cleaned would require, at a minimum, one PQ run to confirm that cleaning performance is consistent with the three original PQ runs.

REVALIDATION

Revalidation is probably not the best term to use for cleaning validation. A better term would be something like *validation confirmation*. Whatever term is used, revalidation refers to the process by which, on a regular basis (as defined in the cleaning validation master plan), a formal evaluation is made of the validity of the cleaning validation previously done. This evaluation is performed to determine whether the original validation work is still applicable to the cleaning process as it is now performed. There are basically two cases of revalidation, with a further subdivision in one of the cases.

Revalidation with a Significant Change

The simplest case is a significant change made in the cleaning process. This significant change may be determined as part of the change control process. For example, if the cleaning agent used is discontinued by the cleaning agent supplier (perhaps because it is primarily marketed for nonvalidated applications) and a replacement must be selected, this usually involves a significant change. Three PQ runs should then be performed with the new cleaning process to "revalidate" the process (in which case one is not really revalidating the old process but rather validating for the first time a *new* process). If the significant change involved equipment, then IQ/OQ on the equipment should be performed followed by the three PQ runs.

Revalidation on a Regular Basis

The second case of revalidation involves evaluation of the initial validation on a regular basis to determine if the initial validation is still valid. This revalidation is usually done on a regular basis, such as every two years. This frequency should be specified in the cleaning validation master plan. On the specified frequency, all data related to the cleaning process should be evaluated. This may include the following information:

1. *All change control done on the cleaning process:* This is evaluated because each change considered in isolation may not be sufficient to conclude that the cleaned process has been significantly changed. However, the accumulation of small changes may sufficiently change the process such that it is a significantly changed process. In such a case, it may be adequate to perform only one PQ run provided the accumulation of small changes is the only reason for questioning whether the initial validation is still applicable.

2. *All change control on the manufacturing process of the product cleaned:* This is performed to evaluate whether the accumulation of small manufacturing changes, each of which in itself is not adequate to determine that the nature and location of cleaned

residues have significantly changed, may cause one to question the applicability of the original validation. In such a case, it may be adequate to perform only one PQ run.

3. *All monitoring data on the cleaning process:* This data are evaluated to give information on the consistency of the process as determined by the monitoring data. The purpose of the monitoring data is to give information that might be an early indication of possible process changes. Monitoring data that are appropriately selected and consistent with the data obtained in the three PQ runs are suggestive of a consistent cleaning process. If monitoring indicates that alert or action levels are exceeded, then this may not be significant for revalidation purposes if an assignable cause is determined and corrections have been made. If there is no assignable cause, and if the system "corrects itself," then there may be reason to question the consistency of the cleaning process. One response to such an occurrence is to increase the level of monitoring (in terms of the number of samples, the frequency of sampling, or increasing the number of tests performed). Consistent data with increased monitoring may be adequate. An alternative approach is to consider performing one PQ run to reconfirm the original validation results.

4. *All QC (quality control) data on products made subsequent to the cleaning process:* If QC data on the lots of products made in the cleaned equipment suggest problems that

may be related to the cleaning process, then it may be appropriate to perform one or more PQ runs to reconfirm the original validation.

5. *All QC data on the lots of cleaned product:* If QC data on lots of cleaned product suggest problems that may affect the nature or location of residues to be cleaned, then it is appropriate to perform one or more PQ runs to reconfirm the original validation.

6. *The original report on the initial cleaning validation:* This should be evaluated to determine whether any of the data from items 1–5 suggest any possible reasons for concluding that the cleaning process is sufficiently different and/or out of control, such that further evaluation is necessary.

It should be noted that before a full revalidation (three PQ runs) is performed, it is usually appropriate to investigate the cleaning process to determine what changes can be made to improve the consistency of that process. The investigation of items 1–6 above should be documented, with a conclusion that either the process is still under control and the original validation is still applicable or that additional work is necessary to confirm the consistency of the process. Following that additional work, it may then be appropriate to conclude that the process is revalidated.

REFERENCES

1.   FDA. 1987. *Guideline on general principles of process validation.* Rockville, Md., USA: Food and Drug Administration, Center for Drug Evaluation and Research, pp. 21–23.

2.   FDA. 1993. *Guide to inspections of validation of cleaning processes.* Rockville, Md., USA: Food and Drug Administration, Office of Regulatory Affairs.

3.   FDA. 1999. *Human drug cGMP notes,* vol. 7, no. 1. Rockville, Md., USA: Food and Drug Administration, Center for Drug Evaluation and Research.

13

Special Topics
in Cleaning Validation

The focus of cleaning validation has, for the most part, been in the finished drug manufacturing segment, since the finished drug product is the product that most directly and immediately touches the patient. However, certain segments of pharmaceutical manufacturing (including segments of finished drug manufacture) have unique challenges because of the nature of the manufacturing or the product use. This chapter covers special issues in cleaning or cleaning validation for several specific industries, including biotechnology, APIs (active pharmaceutical ingredients), manufactured in bulk form contract drug manufacture, clinical trial drug manufacture, and in vitro diagnostic manufacture. Each of these will be discussed below.

BIOTECH MANUFACTURING

There are several issues that are unique to biotech manufacturing. One relates to the analytical procedures used to measure for residues of the active drug. Because many of the products from biotech manufacturing are proteins and are usually cleaned with either caustic alone, with caustic and bleach (sodium hypochlorite), or with formulated alkaline cleaners in hot, aqueous systems, the active protein itself is usually denatured during the cleaning process. Therefore, any technique such as ELISA (enzyme-linked immunosorbent assay), which is valid for measuring the active in the finished drug, is virtually useless for measuring of any residues of that active after the cleaning process. If residues of the active itself are left after the cleaning process, it is usually due to a gross cleaning failure, such as lack of contact of the cleaning solution with the residues to be cleaned (perhaps caused by a clogged spray ball). For this reason, the usual approach for residue detection in the biotech industry is to establish limits for the drug active on the cleaned surface but to measure that drug active indirectly by measuring the total organic carbon (TOC) levels [1,2]. The TOC value is then calculated as if all of the carbon was due to, for example, an active protein. If that calculated protein level is below the established limit for the protein, then the cleaning has been adequate (at least as far as the active protein is concerned). One possible objection to this approach is that if one knows the active protein should not be present (and cannot be present if cleaning is performed correctly), why even set a limit for the protein? Isn't it possible that a degradation product might be of more concern and that a degradation product should be targeted? If one is aware of degradation products that are of special concern (perhaps from a toxicological perspective), then these residues should be targeted. However, in most cases, the proteins are so randomly degraded that it would be difficult to single out certain chemical species to target. In any case, those species would be present at levels no higher than the estimated level of the active drug (assuming the active drug and any degradation products would have roughly the same carbon content). The issue of "what about . . . ?" is one that can be raised in this situation; however, unless there is a good reason to believe a degradation product presents a special risk, it is an issue that has not been seen as a significant risk.

A second issue in biotech manufacture relates to the fact that manufacturers often will attempt to sterilize or decontaminate the equipment before it is cleaned. This is done to kill all microorganisms used in the manufacturing process. If this is done by steam, then this process may result in the proteinaceous material being denatured and/or dried onto equipment surfaces. This results in residues that are much more difficult to clean from the equipment surfaces. For this reason, it is not uncommon in such a situation to employ an oxidizing agent, such as bleach (sodium hypochlorite), to assist in the oxidation and breakdown of the denatured protein residues. The use of bleach requires special controls in order to minimize the effects of chloride ions on the passivated layer of stainless steel equipment.

The third issue related to biotech processing is the concern for microbial control. The main reason for this is not safety of the product but rather manufacturing productivity, efficiency, and/or quality. The presence of a foreign bacterial strain in a fermentation process, for example, may significantly affect the yield or quality of the product manufactured. For this reason, microbial issues are significant for biotech manufacturing.

API MANUFACTURE

There are several issues unique to the manufacture of APIs [3]. The first relates to how residue limits are established in API manufacture. The issues with API manufacture are threefold:

1. Cleaning may be performed at intermediate process steps, and any residues from those cleaning steps might be removed in subsequent processing steps.

2. The level of any residue from earlier cleaning processes present in the final API must be evaluated in light of the effect it has on the *finished* drug product using that API.

3. The effect of a final cleaning process after the API is manufactured must be evaluated in light of possible contamination of the next API manufactured in that equipment.

Any limit established for an intermediate cleaning step should be established in light of how those levels contribute to the cleaning residue levels in the final API. If it can be demonstrated that those residues (intermediates or cleaning agents) are effectively removed from the API during subsequent processing steps (such as recrystallization or chromatographic separation), such that the residue level in the final API is consistent (regardless of how much residue is present at any intermediate step), then residue limits can be established based on process capability. This demonstration that residue levels in the final API are independent of levels present at intermediate steps could be shown with lab studies but should be confirmed with pilot- or full-scale manufacture. If this is the case, then the manufacturer may want to consider these intermediate cleaning steps as *not* critical; hence, the cleaning does not need to be validated. If this approach is taken, then this option should be addressed in the cleaning validation master plan, and any specific decision as to criticality should be clearly documented. It should be noted that, in this case, the final process step is clearly a critical process requiring process validation.

In addition to the effects of the residue(s) on the final API, one should also address the effects of cleaning residues as they affect process steps. In chemical reactions or separations, cleaning residues *may* affect the purity or yield of the API. This has to be established based on process capability.

This discussion does not mean that a cleaning process in an intermediate process step is not important. Rather, it reflects the fact that in such an established process, cleaning residues have usually been indirectly evaluated as part of process capability.

There are two possible sources of cleaning process residues in an API: (1) any possible residue due to intermediate cleaning steps, and (2) any cleaning process residue that is present in the manufacturing equipment prior to any step of manufacture. Both sources should be addressed in considering possible residues. The effect of each subsequent process step in removing those residues should be considered. Whatever their source, residue levels in the final API (the bulk drug before it is released for manufacture into a finished drug product) should be based on the effect that the residue would have in any finished drug product made from that API.

For example, if the maximum permitted level (based on pharmacological or toxicological data, product dosing, and a safety factor) of a specific residue in the finished drug product is 5 ppm, and the level that the API is present in the finished drug product is 0.1 percent, then the maximum amount allowable in the API (R_{API}) can be calculated from the maximum calculated for the drug product ($R_{product}$) as

$$R_{API} = \frac{(R_{product})(100)}{\text{percent active}} = \frac{(5 \text{ ppm})(100)}{0.1} = 5,000 \text{ ppm}$$

In other words, if the API is present at 0.1 percent in the drug product, and the limit of the specific residue (that is present in the API) is established to be 5 ppm in the drug product, then one could produce the API with 5,000 ppm of that specific residue, which is still acceptable for use in the finished drug product.

This assumes, of course, that the API is the only source of that specific residue that could be in the drug product. It may also be appropriate to establish an alternative maximum limit for any cleaning process residue in an API at 0.1 percent (or 1,000 ppm), consistent with the limit for impurities in APIs that do not have to be characterized. In other words, if the calculated limit is 500 ppm, then that limit should be used for validation purposes. If the calculated limit is 5,000 ppm, then one should default to the 1,000 ppm level.

The final issue is that the effect of any residues remaining after cleaning in the final API manufacture must be considered in light of what effect that residue might have if it contaminates the subsequently manufactured API. In turn, the effect in that subsequently manufactured API must be considered in light of the effects of that residue in the finished drug product manufactured with that subsequent API. If the goal is to identify the surface contamination limit in $\mu g/cm^2$, then this can be mathematically expressed as

$$L_{API} = \frac{(R_{API})(BS_{API})(1,000)}{SA}$$

where L_{API} is the residue limit in the manufacturing vessel in $\mu g/cm^2$, R_{API} is the residue limit in subsequently produced API in ppm (or $\mu g/g$), BS_{API} is the batch size of the subsequently manufactured API in kg, and SA is the shared product surface area of equipment in cm^2.

In this equation, 1,000 represents a conversion factor from kilograms to grams. The limit R_{API} can be established as discussed in the previous section. For most cases (unless one is dealing with extremely potent drugs), such a calculation will result in limits significantly above the commonly used visual standard of 4 $\mu g/cm^2$. On an individual basis, the company must decide whether visually clean then becomes the only acceptance criterion for cleaning validation purposes. This may be done only if *all critical* surfaces can be evaluated for visual cleanliness.

The issue of residue limits for cleaning in API manufacture is more complex because it is at least one step removed from the final drug product that will be utilized by the patient. While the Food and Drug Administration (FDA) has required the validation of cleaning of APIs (indeed, one of the initial incidents that started the increased concern over cleaning involved the manufacture of an API), the extent of work required for such validation may be considerably less than required for finished drug manufacture. This is due to the fact that cleaning may be less critical at this stage of manufacture. This may be demonstrated by the effects of purification in subsequent manufacturing steps of the API or by the extremely high calculated residue limits if sound scientific principles are applied to such situations. However, regardless of the extent of validation performed, there is clearly a need that cleaning be appropriately standardized and consistently controlled in the manufacture of APIs.

CONTRACT MANUFACTURE

The main concern with cleaning validation in a contract manufacturing setting relates to the issue of setting residue limits. If residue limits for the cleaning of one product are based in part on the nature of the subsequently manufactured drug product, that subsequently manufactured drug product must be known at the time of cleaning validation. However, the identity of the subsequently manufactured product is often not known in a contract manufacturing situation. Two approaches will be given for this issue: one from the perspective of the contract manufacturer and another from the perspective of the contracting pharmaceutical company.

Contract manufacturers have two options as they approach the issue of documenting the acceptability of the cleaning process. One option is to modify the traditional validation procedure in terms of how residue limits are handled. For the following discussion, product A is the established product for which cleaning validation has been or will be done. Product B is the "new" product being considered for manufacture on the same equipment. In this approach, as with other cleaning validation protocols, residue limits for the cleaning of product A are established based on expected subsequently manufactured products. The decision of what is "expected" can be based on what is currently manufactured at that site in the same equipment and/or what is known about developmental products that might be manufactured in the future. Cleaning validation of product A is then performed with that limit. Then (and this is most important) every time a new product, such as product B, is being considered for manufacture on that same equipment, the contract manufacturer must reevaluate the previously used residue limit for cleaning after product A and determine whether a calculated residue limit for the cleaned equipment is below or above that limit if product B is to be the subsequently manufactured product. If the limit based on product B being the subsequent product is at or above the limit used for the cleaning validation protocol, then product B can be made subsequent to product A in the cleaned equipment. This approach is in part no different from the approach taken in multiuse equipment when a new product is proposed for manufacture on that equipment when cleaning validation for other products has previously been done (although it differs in that grouping strategies are more difficult for a contract manufacturer). If the limit for cleaning of product A when product B is the subsequently manufactured product is below that used in the cleaning validation protocol, then the contract manufacture has three options to pursue based on probable relative cost: (1) repeat the cleaning validation on product A at the lower limit, (2) perform cleaning verification (see Chapter 1) on the manufacture of product A when product B is the subsequent product, and (3) restrict the manufacturing sequence such that product B is never made subsequent to product A.

The second situation is does not involve validation at all but rather treats each cleaning situation as unique, which requires a strict approach of cleaning verification on each and every cleaning process (or at least on the final cleaning involving product changeover). While this would address the issues of dealing with "unknown" subsequent products, it has a number of disadvantages. One might think one could avoid the costs of validation by implementing a verification only program. However, the reality is such that over multiple cleaning events, the costs of testing in a verification program will be more expensive than the costs of testing in a validation program. The extensive testing in a validation framework may stop after the three process qualification (PQ) runs. In a verification framework, more testing is generally done on each cleaning event and continues on each and every cleaning event. In addition, there may be issues relating to the use of cleaned equipment pending the results of analytical testing, reducing the availability of that process equipment for use. These are business decisions involving equally justified scientific options. It should be noted that the option of just doing cleaning verification in a contract manufacturing situation has not (to my knowledge) been tested with the FDA, because the option is generally not an economical one.

Pharmaceutical companies who use contract manufacturing facilities have two concerns when addressing cleaning and cleaning validation. The first consideration is what cleaning is done *before* their product is manufactured in the contract facility. This is a concern because they obviously do not want their products manufactured on equipment with residues that might unacceptably contaminate their product. One issue involved in this is a knowledge of what products are made before their product, the cleaning process and cleaning validation performed on that previously manufactured product, and the acceptance criteria established for that cleaning process. In addition, the pharmaceutical company may want to require that only certain products (for which they have information) be manufactured immediately before their product is manufactured. Such information exchange (essentially between two companies using the same contract facility) can usually be conducted under a secrecy agreement; such an exchange should be agreeable because in most cases the exchange is mutual (if company A and company B both use the same contract manufacturer and the products of either company can be made in any manufacturing sequence, then company A will want to know about the cleaning done on the products of company B, and company B will want to know about the cleaning done on the products of company A). Optionally, the evaluation of the acceptability of the cleaning validation of the previously manufactured products can be done by a qualified third party acceptable to both companies and to the contract manufacturer.

The second concern of a company contracting the manufacture of its products is that appropriate cleaning and validation are done *after* the manufacture of that company's products. The rationale for this is that cleaning validation may be required by regulatory authorities. The second reason is that the company does not want its products contaminating a subsequently manufactured product of another company because of the damage it might do to the consumer of that subsequently manufactured product and the adverse publicity and/or lawsuits related to such contamination. In essence, these cleaning concerns are the same as with the manufacture of any pharmaceutical product. Is the equipment suitably cleaned so that the product can be made? Has the equipment been suitably cleaned after the manufacture of the product so that any subsequently manufactured product is not contaminated? In contract manufacture, these issues are highlighted because of the presence of two (or more) additional parties: the contract manufacturer and any other company also contracting products to be manufactured on the same equipment.

CLINICAL MANUFACTURE

The issues involved in clinical manufacture are similar to those involved in contract manufacture but with some additional complications. For clinical manufacture, the subsequent product(s) made in a given piece of equipment may be unknown at the time cleaning is done. In addition, the same product may not be made on the same equipment using the same manufacturing process for the requisite three runs required for validation purposes. Fortunately, the FDA has commented on the requirement for cleaning for clinical manufacture [4]. Basically, their response was to acknowledge that cleaning validation may not be required in such situations; however, cleaning verification would be required. This means that a cleaning SOP (Standard Operating Procedure) must be in place, and, at the end of every cleaning process, the equipment must be evaluated by appropriate analytical procedures to determine the levels of residues. The determination of the acceptability of those residues cannot be finalized until the subsequently manufactured product in that same equipment is identified. At that time, an evaluation is done using the principles given in Chapter 8 to determine whether the residues actually measured are acceptable in light of that subsequently manufactured product. This cleaning verification should be appropriately documented.

The second issue with clinical manufacture is that inadequate information may be available to determine the minimum dose of the cleaned active.

Since the minimum dose of the cleaned active is used in setting appropriate acceptance criteria, how does one set limits, particularly in the early stages of clinical trials? This is usually accomplished by basing limits on toxicity information (such as an LD_{50}) and establishing an estimated acceptable daily intake (ADI) based on that toxicity information using accepted guidelines [5]. Certainly any other available relevant information should be used in establishing this limit. Additional safety factors may also be considered in establishing acceptance limits.

IN VITRO DIAGNOSTICS

While the concept of evaluating the effect of the residue on subsequently manufactured products is valid when applied to an in vitro diagnostic (IVD) product [6], the concept of looking for pharmacological effects is not. In place of a pharmacological effect, one can consider evaluating residues for potential effects on the functionality of any subsequently manufactured IVD product. Limits should be placed on residues to achieve levels that are below the level at which those residues measurably affect the performance of the subsequently manufactured IVD product.

Residues that can be considered as potentially contaminating include the previously manufactured IVD or the cleaning agent. Effects on the functionality of the subsequently manufactured product may be determined by spiking experiments. For example, in laboratory studies, fixed amounts of target residues may be deliberately added to the subsequently manufactured product at levels of 1, 5, 20, and 100 ppm (or whatever may be appropriate, depending on the residue and the IVD product). Then the performance of that spiked IVD product may be evaluated using positive and negative controls. It is desirable but not necessary to establish a level of contaminant that does have a demonstrated effect on the performance of the subsequently manufactured IVD product. In evaluating the test performance of the subsequently manufactured IVD product, it may also be appropriate to evaluate test performance after accelerated stability testing of the spiked IVD product. The acceptable contamination level is the maximum level tested that has no effect on test performance. A safety factor of 2–10 may be added to this no effect level. For a multiproduct facility in which products may be manufactured in any order, it may be necessary to determine residue limits for cleaning purposes with each of the possible subsequently manufactured products. The actual acceptance limit chosen should be the lowest among those calculated for the possible subsequent products.

The important consideration in all of these situations is that sound scientific principles are applied that are relevant to the specific situation. The complicating factor of cleaning validation in many of these situations is that it does require consideration of the subsequently manufactured product, and there may be limited information on that subsequently manufactured product.

REFERENCES

1.   Baffi, R., G. Dolch, R. Garnick, Y. F. Huang, B. Mar, D. Matsuhiro, B. Niepelt, C. Parra, and M. Stephan. 1991. A total organic carbon analysis method for validating cleaning between products in biopharmaceutical manufacturing. *Journal of Parenteral Science & Technology* 45 (1):13–19.

2. Jenkins, K. M., et al. 1996. Application of total organic carbon analysis to cleaning validation. *Journal of Parenteral Science and Technology* 50 (1):6–15.

3. For a good introduction to cleaning validation for APIs, see: PhRMA guideline for the validation of cleaning processes for bulk pharmaceutical chemicals. 1997. *Pharmaceutical Technology* 21 (9):56–73.

4. FDA. 1997. *Human drug cGMP notes*, vol. 3, no. 5, Rockville, Md., USA: Food and Drug Administration, Center for Drug Evaluation and Research.

5. Conine, D. L., B. D. Navmann, and L. H. Hecker. 1992. Setting health-based residue limits for contaminants in pharmaceuticals and medical devices. *Quality Assurance: Good Practice, Regulation and Law* 1 (3):171–180.

6. FDA. 1994. *Guidelines for the manufacture of in vitro diagnostic products.* Rockville, Md., USA: Food and Drug Administration, Center for Devices and Radiological Health.

14

FDA Expectations

The primary expectation of the Food and Drug Administration (FDA) is that for pharmaceutical manufacturing processes, including finished drugs and active pharmaceutical ingredients (APIs) for both human and animal use, any *critical* cleaning process must be validated. Critical cleaning processes are those that involve surfaces that, if cleaned inadequately, have a reasonable probability of contaminating the subsequently manufactured product. For the most part, this involves product contact surfaces, surfaces that the subsequently manufactured product actually contacts, allowing for the possibility of direct transfer of the residues from the surface to that product. Other nonproduct-contact surfaces may also be considered critical, provided there is evidence (or a reasonable scientific judgment) that residues on those surfaces may transfer to the manufactured product. For nonproduct-contact surfaces, the issue of validation has to be decided on the specifics of the situation. The definition of critical cleaning processes is not explicitly stated in the FDA cleaning validation guidance document. However, it is one that appears to be assumed and is consistent with the requirement in the guidance document on process validation that focuses on manufacturing processes [1,2].

A second expectation of the FDA is that each manufacturer analyzes and understands its own cleaning processes and establishes acceptance criteria based on good scientific principles. FDA investigators are very good at identifying unacceptable practices; however, they are trained not to prescribe acceptable practices. An FDA investigator does not have the time to fully understand any individual cleaning situation; the manufacturer should fully understand its own cleaning situation and should be in a position to establish acceptable processes. For example, if a manufacturer were to perform cleaning validation using only 10 ppm as an arbitrarily set acceptance criterion for all residues, it is likely that an FDA investigator will challenge the basis for selecting of that residue limit. The investigator may refer the manufacturer to the FDA cleaning validation guidance document, but the primary prescription will be that the residue limits have a scientific justification [3].

The rest of this chapter will cover validation issues described in the FDA guidance document, *Guide to Inspections of Validation of Cleaning Processes*, which is included as Appendix A of this book [4]. This guide was issued in July 1993, so it reflects the FDA's thinking at that time. As with other guidance documents, it is designed for FDA investigators, not manufacturers. However, manufacturers should understand the document because the issues raised in it address the minimum expectations from the FDA in an inspection of cleaning validation. The guidance document does not cover every issue that an investigator may address, nor does the document forbid other methods of achieving the same end. However, any manufacturer using alternative methods should be prepared to give a valid scientific justification of the acceptability of those methods.

One other issue about the guidance document is significant—the statement that it "is intended to cover equipment cleaning for chemical residues only." As discussed in Chapter 11, there has been more and more focus on microbial residues for cleaning validation purposes. Despite the guidance document disclaimer, the principles covered in this document can be, for the most part, directly applied to microbial issues. The major exception is how residue limits are established for microbial residues. Furthermore, despite that disclaimer, the document actually addresses microbial issues by discussing the importance of bioburden reduction in the cleaning process.

What follows are a recap and discussion of topics in the 1993 guidance document. The headings below are similar to headings in that guidance document.

GENERAL REQUIREMENTS

Since cleaning validation involves the validation of a cleaning process, the FDA expects that manufacturers will have written procedures or SOPs (Standard Operating Procedures) in place that detail the cleaning process. The key word here is "detail"; the SOPs have to be in sufficient detail to describe the process. This is further discussed later in this chapter. SOPs may be different for different residues on the same equipment, but such distinctions in applicability must be clearly specified. Also, cleaning between batches of the same product may be different from cleaning at product changeovers; such distinctions in applicability must be clearly specified.

The document also specifies that the FDA expects written procedures for how cleaning validation is performed. These procedures should address responsibility for performing and approving the study, how acceptance criteria are determined, and when revalidation is necessary. Most of these items would be covered in a cleaning validation master plan. It should be noted that this is not a requirement for a cleaning validation master plan, because there are elements of the master plan that are not specified in this document. Nonetheless, a cleaning validation master plan is probably the most efficient way to meet these regulatory requirements as well as to accomplish validated cleaning processes in a timely, consistent, and defendable manner.

Under this section, the FDA also states that it expects written validation protocols prepared in advance, addressing issues such as sampling and analytical methods for residues, including the sensitivity of these methods. Sampling and analytical methods are at the heart of the validity of any cleaning validation protocol, so one can expect special attention to these issues in any inspection. The issue of sensitivity, which as a practical matter is to be understood as limit of detection (LOD) or limit of quantitation (LOQ), is simply that the analytical method must be able to detect the target residues at least down to the acceptance criterion in the analytical sample.

Finally, the FDA states that the validation study must be conducted in accordance with the protocol, and the final validation report, stating whether or not the cleaning process is validated, must be approved by management. These general requirements are consistent with the FDA's expectations for the validation of other manufacturing processes.

EVALUATION OF CLEANING VALIDATION

The FDA basically says that manufacturers need to understand their cleaning processes before jumping in with elaborate sampling and analytical schemes. This is generally done through prequalification testing (see Chapter 6). The FDA suggests that this understanding may also allow manufacturers to design less costly procedures, although it should be clearly recognized that the objective of an FDA inspection is not that the cleaning process be optimized from an economic perspective but rather that it is adequate for its intended purpose.

The FDA also states that "ideally, a piece of equipment or system will have one process for cleaning." However, the context clearly recognizes that this will depend on the individual situation. This does not preclude having different cleaning processes on the same piece of equipment for different manufactured products. For example, the manufacture of product A on a tablet press may require cleaning process X, while the manufacture of product B on the same tablet press may require cleaning process Y. What should be avoided is the specification of alternative cleaning processes for the same manufactured product. For example, if one specified for the manufacture of product B that either cleaning process Y or cleaning process Z is acceptable, what are the consequences? The first consequence is that both cleaning processes (Y and Z) will have to be validated for that manufactured product. The second consequence is that (despite the SOP), the wrong cleaning process might be used. For example, if the processes are manual, and if process Y specifies that detergent M is used at 1 percent at 40°C for 5 minutes and process Z specifies that detergent N is used at 1.5 percent at 45°C for 10 minutes, the specification of the alternatives may result in detergent N being used at 1 percent at 45°C for 5 minutes, resulting in questionable results.

A final significant issue in this section is that a cleaning process used on equipment between batches of the same product need only meet an acceptance criterion of visibly clean. Furthermore, such a cleaning process does not require validation. Although not stated, the rationale for the latter statement is that cleaning between lots of the same product is not a critical cleaning process, because the worst that can happen is that residues of the previous lot carry over to the next lot. While this may compromise lot integrity, it is an economic risk, not a safety one; therefore, validation is not required. Notwithstanding this statement, many firms validate cleaning between lots of the same product. This occurs because (1) they want to have one quality standard for all products manufactured in the facility; (2) they realize that the validation process, if the main acceptance criterion is visibly clean, is considerably simplified; and (3) validation of cleaning helps support an assertion of lot integrity for batches in the same campaign. It should also be noted that even though visibly clean may be adequate, manufacturers are responsible for understanding their cleaning process. If there is evidence that other criteria are appropriate, such as the presence of degradation products, the possibility of cleaning agents contaminating the products, or microbial contamination, the use of a criterion other than visibly clean should be considered. Finally, this issue is sometimes interpreted as dedicated equipment does not require validation. In this case, the term *dedicated* should be further defined. Some manufacturers would consider all equipment used only for topical products as dedicated. This is *not* the intention in this section. Dedicated sometimes refers to equipment used only for manufacturing lots of the same product. However, here the FDA guidance document goes beyond dedicated equipment and applies also to nondedicated equipment, for cleaning between successive lots of the same product. Of course, the expectation at changeover in such equipment is that the cleaning be validated.

Equipment Design

The guidance document instructs the investigator to examine the equipment design. This includes both the equipment cleaned and, in the case of automated processes such as CIP (clean-in-place) systems, the cleaning equipment. One specific issue related to the cleaning systems is the expression of a preference for "sanitary type piping without ball valves." Sanitary piping refers to the food industry standards for sanitary piping [5], including such items as avoidance of threaded fittings (a good hiding place for residues), weld quality, and pitch of piping. Sanitary items are preferred and should be specified in an initial design. However, in dealing with the cleaning validation on established equipment, these issues are not restrictions but require special attention in the design of the cleaning process and in the selection of sampling points for the analysis of residues. For example, ball valves in a system may require disassembly for cleaning. Because of their difficulty of cleaning, they may be selected as one of the locations where residues are sampled.

The FDA also suggests that investigators look at the training records of operators for any difficult cleaning processes, such as those involving ball valves. It is also suggested that the tagging and identification of valves and piping be appropriate for correct execution of the cleaning process. Finally, the FDA suggests looking for an SOP step that identifies the length of time that dirty equipment may stand idle before the cleaning process is begun. The concern here is that as the residues dry onto the equipment surfaces, they may be more difficult to remove. The usual industry approach to this issue has been to specify a maximum time between the end of processing and the beginning of cleaning and to include that maximum time in at least one of the three Process Qualification (PQ) runs for validation purposes.

While the introduction to the guidance document specifies that the document applies only to chemical residues, this section also specifies that there should be some evidence that cleaning and storage do not allow microbial contamination. The main issue here is the drying of equipment for storage. The document states that equipment should be dried before storage, "and under no circumstances should stagnant water be allowed to remain in equipment subsequent to cleaning operations." The focus of this is equipment storage; if the equipment is immediately used, one would not expect microbial proliferation on equipment that was not dried. The key issue is, "What constitutes storage?" Most people would agree that using the equipment one hour after cleaning does not involve storage, and hence does not require special attention. Most people also would agree that cleaned equipment unused for three days probably qualifies as being "in storage" and should be dried. However, there may be gray areas in between where scientists would disagree. The solution in these cases is to develop data showing that the cleaning process used and the delay before next manufacture do not cause microbial proliferation.

The final issue discussed in this section is the effect of the cleaning process on bioburden for equipment that is subsequently sterilized or sanitized. It is well known that bioburden should be controlled in the validation of any sterilization or sanitation process. Therefore, an evaluation of bioburden after cleaning should be considered in such cases. Also, steam sterilization processes have no significant effect on removal or inactivation of endotoxins (pyrogens). Cleaning processes may also be evaluated for their contribution to the removal of endotoxins, particularly for equipment used for aseptic processing.

Cleaning Process Written

How much detail is required in the cleaning SOP? There are no specific rules; complex cleaning procedures require more documentation, and simple procedures may require much less documentation. For more complex procedures, this document suggests that critical steps in the cleaning process be identified. While not specified, it is a reasonable expectation that critical steps require double sign-off. In addition, the guidance document suggests that procedures requiring operators to perform complex manipulations would also require more documentation in the SOP.

Analytical Methods

The sensitivity and specificity of the analytical method must be determined. Sensitivity refers to the LOD or LOQ, which should preferably be below the acceptance criterion for the analytical sample. Although nothing more is said about specificity, this is sometimes misinterpreted to mean that only a specific method is acceptable. This is clearly not the case, and the reader is referred to Chapter 9 for a more detailed discussion of this issue.

Finally, the section specifies that the analytical method should be evaluated with the sampling procedure to determine the percent recovery from equipment surfaces. This recovery value is then used to modify any analytical results obtained from sampling surfaces (see Chapter 10).

Sampling

There are two sampling procedures that are acceptable: direct surface sampling (which is called swab sampling in this book in Chapter 10) and rinse sampling. The advantages of swabs are that the worst-case locations (hardest to clean) can be targeted. A second advantage is that insoluble residues can be physically removed for subsequent analysis (of course, those insoluble residues may need to be subsequently solubilized prior to actual analysis). This document cautions that interferences from the sampling process (namely the swab itself) must be considered.

The advantages of rinse sampling are twofold: (a) the sampling of a larger surface area and (b) the sampling of inaccessible locations (those locations inaccessible by swabs). Here the FDA gives their "dirty pot" analogy. The main disadvantage of rinse sampling is that the residue may not be soluble in the rinse water, and, therefore, analysis of the rinse water may not be appropriate. As discussed in Chapter 10, this can be addressed with a recovery procedure with the rinse sampling just as one would for swab sampling. A second caution on rinse sampling is that analysis of the rinse sample should include a direct measurement of the targeted residue. It is not adequate to test the rinse sample for compendial specifications (such as for USP [U.S. Pharmacopeia] Purified Water) and conclude that the equipment is acceptably clean; an analytical test that correlates with the targeted residue must be utilized.

Routine Production In-Process Control

In-process control involves monitoring after the cleaning process is validated. The main issue here is that whatever method is used for monitoring in some way correlates with the equipment condition ("clean" or "not cleaned"). For this reason, the FDA suggests that any indirect monitoring procedure (such as conductivity) is tested to confirm that uncleaned equipment (or unacceptably cleaned equipment) give a not acceptable result by the indirect monitoring procedure.

Establishment of Limits

The FDA clearly states that it does not intend to set acceptance criteria for manufacturers. However, it expects manufacturers' acceptance criteria to be "logical . . . , practical, achievable, and verifiable." The emphasis is that the limits are scientifically justifiable. There is a reference to the Fourman and Mullen paper limits of 10 ppm, levels of 1/1,000 of the normal therapeutic dose, and visibly clean. Because of this reference or the scientific soundness of that approach, such an approach or a modification of it is used by most manufacturers for acceptance limits (see Chapter 8).

In addition, for API manufacture where there may be partial reactants and by-products involved, focusing on the principal reactant may be inadequate. The FDA suggests that supplementary techniques such as thin layer chromatography may be appropriate.

Other Issues

Placebo Product

The issue of placebo sampling is discussed in detail in Chapter 10. The major concerns expressed in this guidance document about placebo sampling are the issues of the analytical power being reduced by sampling in the placebo matrix and the issue of nonuniform contamination of the subsequent product. For this reason, the FDA states that placebo testing is acceptable only if used in conjunction with swab or rinse testing.

Detergent

There are several issues brought up with the use of detergents. The first relates to the detergent composition. Since the pharmaceutical manufacturer is responsible for the evaluation of possible residues, that manufacturer must know the composition of the detergent to make a proper evaluation. The second issue for detergents is that, because detergents are "not a part of the manufacturing process," residues of those detergents should be easily removable and "no" or "very low" levels should remain after the cleaning process. While one clearly understands the FDA intent here, the rationale that it is not a part of the manufacturing process is inconsistent with the principle that cleaning is part of the manufacturing process and therefore must be validated. Regardless of the justification, the FDA is correct in stating that the detergent selected should be freely rinsing and should therefore be present in very low levels.

Test Until Clean

The FDA addresses the issue of retesting (i.e., test until clean) when unacceptable results are obtained. The FDA clearly states that repeated retesting shows that the cleaning process is not consistent and hence not validated. It should be noted in this regard that the issue of how out-of-specification (OOS) test results are handled is currently being addressed by the FDA [6,7], and OOS cleaning validation or monitoring test data should be handed in the same manner.

SUMMARY

While this chapter covers some of the published FDA expectations in considering cleaning processes and cleaning validation, it certainly is not exhaustive. Just as cleaning validation became a high profile issue with the FDA because of specific events that involved product contaminated due to uncontrolled cleaning processes, one can expect that new high profile issues related to cleaning validation may arise due to observed, but unexpected, problems with validated cleaning as it is now done.

REFERENCES

1. FDA. 1987. *Guideline on general principles of process validation.* Rockville, Md., USA: Food and Drug Administration, Center for Drug Evaluation and Research.

2. FDA. 1996. Current good manufacturing practice; Proposed amendment of certain requirements for finished pharmaceuticals. *Federal Register* 61:20103.

3. FDA. 1998. *Human drug cGMP notes,* vol. 6, no. 2. Rockville, Md., USA: Food and Drug Administration, Center for Drug Evaluation and Research.

4. FDA. 1993. *Guide to inspections of validation of cleaning processes.* Rockville, Md., USA: Food and Drug Administration, Office of Regulatory Affairs.

5. *3A Accepted Practices.* Ames, Ia., USA: International Association of Milk, Food, and Environmental Sanitarians.

6. FDA. 1998. *Guide to inspections of pharmaceutical quality control laboratories.* Rockville, Md., USA: Food and Drug Administration, Office of Regulatory Affairs.

7. FDA. 1998. *Draft guidance for industry: Investigating out of specification (OOS) test results for pharmaceutical production.* Rockville, Md., USA: Food and Drug Administration, Center for Drug Evaluation and Research.

Appendix A

Guide to Inspections of Validation of Cleaning Processes (July 1993)

Note: This document is reference material for investigators and other FDA personnel. The document does not bind FDA, and does not confer any rights, privileges, benefits, or immunities for or on any person(s).

I. INTRODUCTION

Validation of cleaning procedures has generated considerable discussion since agency documents, including the Inspection Guide for Bulk Pharmaceutical Chemicals and the Biotechnology Inspection Guide, have briefly addressed this issue. These Agency documents clearly establish the expectation that cleaning procedures (processes) be validated.

This guide is designed to establish inspection consistency and uniformity by discussing practices that have been found acceptable (or unacceptable). Simultaneously, one must recognize that for cleaning validation, as with validation of other processes, there may be more than one way to validate a process. In the end, the test of any validation process is whether scientific data shows that the system consistently does as expected and produces a result that consistently meets predetermined specifications.

This guide is intended to cover equipment cleaning for chemical residues only.

II. BACKGROUND

For FDA to require that equipment be clean prior to use is nothing new, the 1963 GMP Regulations (Part 133.4) stated as follows "Equipment *** shall be maintained in a clean and orderly manner ***." A very similar section on equipment cleaning (211.67) was included in the 1978 CGMP regulations. Of course, the main rationale for requiring clean equipment is to prevent contamination or adulteration of drug products. Historically, FDA investigators have looked for gross insanitation due to inadequate cleaning and maintenance of equipment and/or poor dust control systems. Also, historically speaking, FDA was more concerned about the contamination of nonpenicillin drug products with penicillins or the cross-contamination of drug products with potent steroids or hormones. A number of products have been recalled over the past decade due to actual or potential penicillin cross-contamination.

One event which increased FDA awareness of the potential for cross contamination due to inadequate procedures was the 1988 recall of a finished drug product, Cholestyramine Resin USP. The bulk pharmaceutical chemical used to produce the product had become contaminated with low levels of intermediates and degradants from the production of agricultural pesticides. The cross-contamination in that case is believed to have been due to the reuse of recovered solvents. The recovered solvents had been contaminated because of a lack of control over the reuse of solvent drums. Drums that had been used to store recovered solvents from a pesticide production process were later used to store recovered solvents used for the resin manufacturing process. The firm did not have adequate controls over these solvent drums, did not do adequate testing of drummed solvents, and did not have validated cleaning procedures for the drums.

Some shipments of this pesticide contaminated bulk pharmaceutical were supplied to a second facility at a different location for finishing. This resulted in the contamination of the bags used in that facility's fluid bed dryers with pesticide contamination. This in turn led to cross contamination of lots produced at that site, a site where no pesticides were normally produced.

FDA instituted an import alert in 1992 on a foreign bulk pharmaceutical manufacturer which manufactured potent steroid products as well as non-steroidal products using common equipment. This firm was a multi-use bulk pharmaceutical facility. FDA considered the potential for cross-contamination to be significant and to pose a serious health risk to the public. The firm had only recently started a cleaning validation program at the time of the inspection and it was considered inadequate by FDA. One of the reasons it was considered inadequate was that the firm was only looking for evidence of the absence of the previous compound. The firm had evidence, from TLC tests on the rinse water, of the presence of residues of reaction byproducts and degradants from the previous process.

III. GENERAL REQUIREMENTS

FDA expects firms to have written procedures (SOP's) detailing the cleaning processes used for various pieces of equipment. If firms have one cleaning process for cleaning between different batches of the same product and use a different process for cleaning between product changes, we expect the written procedures to address these different scenarios. Similarly, if firms have one process for removing water soluble residues and another process for non-water soluble residues, the written procedure should address both scenarios and make it clear when a given procedure is to be followed. Bulk pharmaceutical firms may decide to dedicate certain equipment for certain chemical manufacturing process steps that produce tarry or gummy residues that are difficult to remove from the equipment. Fluid bed dryer bags are another example of equipment that is difficult to clean and is often dedicated to a specific product. Any residues from the cleaning process itself (detergents, solvents, etc.) also have to be removed from the equipment.

FDA expects firms to have written general procedures on how cleaning processes will be validated.

FDA expects the general validation procedures to address who is responsible for performing and approving the validation study, the acceptance criteria, and when revalidation will be required.

FDA expects firms to prepare specific written validation protocols in advance for the studies to be performed on each manufacturing system or piece of equipment which should address such issues as sampling procedures, and analytical methods to be used including the sensitivity of those methods.

FDA expects firms to conduct the validation studies in accordance with the protocols and to document the results of studies.

FDA expects a final validation report which is approved by management and which states whether or not the cleaning process is valid. The data should support a conclusion that residues have been reduced to an "acceptable level."

IV. EVALUATION OF CLEANING VALIDATION

The first step is to focus on the objective of the validation process, and we have seen that some companies have failed to develop such objectives. It is not unusual to see manufacturers use extensive sampling and testing programs following the cleaning process without ever really evaluating the effectiveness of the steps used to clean the equipment. Several questions need to be addressed when evaluating the cleaning process. For example, at what point does a piece of equipment or system become clean? Does it have to be scrubbed by hand? What is accomplished by hand scrubbing rather than just a solvent wash? How variable are manual cleaning processes from batch to batch and product to product? The answers to these questions are obviously important to the inspection and evaluation of the cleaning process since one must determine the overall effectiveness of the process. Answers to these questions may also identify steps that can be eliminated for more effective measures and result in resource savings for the company.

Determine the number of cleaning processes for each piece of equipment. Ideally, a piece of equipment or system will have one process for cleaning, however this will depend on the products being produced and whether the cleanup occurs between batches of the same product (as in a large campaign) or between batches of different products. When the cleaning process is used only between batches of the same product (or different lots of the same intermediate in a bulk process) the firm need only meet a criteria of, "visibly clean" for the equipment. Such between batch cleaning processes do not require validation.

1. Equipment Design

Examine the design of equipment, particularly in those large systems that may employ semi-automatic or fully automatic clean-in-place (CIP) systems since they represent significant concern. For example, sanitary type piping without ball valves should be used. When such nonsanitary ball valves are used, as is common in the bulk drug industry, the cleaning process is more difficult.

When such systems are identified, it is important that operators performing cleaning operations be aware of problems and have special training in cleaning these systems and valves. Determine whether the cleaning operators have knowledge of these systems and the level of training and experience in cleaning these systems. Also check the written and validated cleaning process to determine if these systems have been properly identified and validated.

In larger systems, such as those employing long transfer lines or piping, check the flow charts and piping diagrams for the identification of valves and written cleaning procedures. Piping and valves should be tagged and easily identifiable by the operator performing the cleaning function. Sometimes, inadequately identified valves, both on prints and physically, have led to incorrect cleaning practices.

Always check for the presence of an often critical element in the documentation of the cleaning processes; identifying and controlling the length of time between the end of processing and each cleaning step. This is especially important for topicals, suspensions, and bulk drug operations. In such operations, the drying of residues will directly affect the efficiency of a cleaning process.

Whether or not CIP systems are used for cleaning of processing equipment, microbiological aspects of equipment cleaning should be considered. This consists largely of preventive measures rather than removal of contamination once it has occurred. There should be some evidence that routine cleaning and storage of equipment does not allow microbial proliferation. For example, equipment should be dried before storage, and under no circumstances should stagnant water be allowed to remain in equipment subsequent to cleaning operations.

Subsequent to the cleaning process, equipment may be subjected to sterilization or sanitization procedures where such equipment is used for sterile processing, or for nonsterile processing where the products may support microbial growth. While such sterilization or sanitization procedures are beyond the scope of this guide, it is important to note that control of the bioburden through adequate cleaning and storage of equipment is important to ensure that subsequent sterilization or sanitization procedures achieve the necessary assurance of sterility. This is also particularly important from the standpoint of the control of pyrogens in sterile processing since equipment sterilization processes may not be adequate to achieve significant inactivation or removal of pyrogens.

2. Cleaning Process Written

Procedure and Documentation

Examine the detail and specificity of the procedure for the (cleaning) process being validated, and the amount of documentation required. We have seen general SOPs, while others use a batch record or log sheet system that requires some type of specific documentation for performing each step. Depending upon the complexity of the system and cleaning process and the ability and training of operators, the amount of documentation necessary for executing various cleaning steps or procedures will vary.

When more complex cleaning procedures are required, it is important to document the critical cleaning steps (for example certain bulk drug synthesis processes). In this regard, specific documentation on the equipment itself which includes information about who cleaned it and when is valuable. However, for relatively simple cleaning operations, the mere documentation that the overall cleaning process was performed might be sufficient.

Other factors such as history of cleaning, residue levels found after cleaning, and variability of test results may also dictate the amount of documentation required. For example, when variable residue levels are detected following cleaning, particularly for a process that is believed to be acceptable, one must establish the effectiveness of the process and operator performance. Appropriate evaluations must be made and when operator performance is deemed a problem, more extensive documentation (guidance) and training may be required.

3. Analytical Methods

Determine the specificity and sensitivity of the analytical method used to detect residuals or contaminants. With advances in analytical technology, residues from the manufacturing and cleaning processes can be detected at very low levels. If levels of contamination or residual are not detected, it does not mean that there is no residual contaminant present after cleaning. It only means that levels of contaminant greater than the sensitivity or detection limit of the analytical method are not present in the sample. The firm should challenge the analytical method in combination with the sampling method(s) used to show that contaminants can be recovered from the equipment surface and at what level, i.e., 50% recovery, 90%, etc. This is necessary before any conclusions can be made based on the sample results. A negative test may also be the result of poor sampling technique (see below).

4. Sampling

There are two general types of sampling that have been found acceptable. The most desirable is the direct method of sampling the surface of the equipment. Another method is the use of rinse solutions.

   a. Direct Surface Sampling—Determine the type of sampling material used and its impact on the test data since the sampling material may interfere with the test. For example, the adhesive used in swabs has been found to interfere with the analysis of samples. Therefore, early in the validation program, it is important to assure that the sampling medium and solvent (used for extraction from the medium) are satisfactory and can be readily used.

Advantages of direct sampling are that areas hardest to clean and which are reasonably accessible can be evaluated, leading to establishing a level of contamination or residue per given surface area. Additionally, residues that are "dried out" or are insoluble can be sampled by physical removal.

b. Rinse Samples—Two advantages of using rinse samples are that a larger surface area may be sampled, and inaccessible systems or ones that cannot be routinely disassembled can be sampled and evaluated.

A disadvantage of rinse samples is that the residue or contaminant may not be soluble or may be physically occluded in the equipment. An analogy that can be used is the "dirty pot." In the evaluation of cleaning of a dirty pot, particularly with dried out residue, one does not look at the rinse water to see that it is clean; one looks at the pot.

Check to see that a direct measurement of the residue or contaminant has been made for the rinse water when it is used to validate the cleaning process. For example, it is not acceptable to simply test rinse water for water quality (does it meet the compendia tests) rather than test it for potential contaminates.

c. Routine Production In-Process Control

Monitoring—Indirect testing, such as conductivity testing, may be of some value for routine monitoring once a cleaning process has been validated. This would be particularly true for the bulk drug substance manufacturer where reactors and centrifuges and piping between such large equipment can be sampled only using rinse solution samples. Any indirect test method must have been shown to correlate with the condition of the equipment. During validation, the firm should document that testing the uncleaned equipment gives a not acceptable result for the indirect test.

V. ESTABLISHMENT OF LIMITS

FDA does not intend to set acceptance specifications or methods for determining whether a cleaning process is validated. It is impractical for FDA to do so due to the wide variation in equipment and products used throughout the bulk and finished dosage form industries. The firm's rationale for the residue limits established should be logical based on the manufacturer's knowledge of the materials involved and be practical, achievable, and verifiable. It is important to define the sensitivity of the analytical methods in order to set reasonable limits. Some limits that have been mentioned by industry representatives in the literature or in presentations include analytical detection levels such as 10 PPM, biological activity levels such as 1/1000 of the normal therapeutic dose, and organoleptic levels such as no visible residue.

Check the manner in which limits are established. Unlike finished pharmaceuticals where the chemical identity of residuals are known (i.e., from actives, inactives, detergents) bulk processes may have partial reactants and unwanted by-products which may never have been chemically identified. In establishing residual limits, it may not be adequate to focus only on the principal reactant since other chemical variations may be more difficult to remove. There are circumstances where TLC screening, in addition to chemical analyses, may be needed. In a bulk process, particularly for very potent chemicals such as some steroids, the issue of by-products needs to be considered if equipment is not dedicated. The objective of the inspection is to ensure that the basis for any limits is scientifically justifiable.

VI. OTHER ISSUES

a. Placebo Product

In order to evaluate and validate cleaning processes some manufacturers have processed a placebo batch in the equipment under essentially the same operating parameters used for processing product. A sample of the placebo batch is then tested for residual contamination. However, we have documented several significant issues that need to be addressed when using placebo product to validate cleaning processes.

One cannot assure that the contaminate will be uniformly distributed throughout the system. For example, if the discharge valve or chute of a blender are contaminated, the contaminant would probably not be uniformly dispersed in the placebo; it would most likely be concentrated in the initial discharge portion of the batch. Additionally, if the contaminant or residue is of a larger particle size, it may not be uniformly dispersed in the placebo.

Some firms have made the assumption that a residual contaminant would be worn off the equipment surface uniformly; this is also an invalid conclusion. Finally, the analytical power may be greatly reduced by dilution of the contaminate. Because of such problems, rinse and/or swab samples should be used in conjunction with the placebo method.

b. Detergent

If a detergent or soap is used for cleaning, determine and consider the difficulty that may arise when attempting to test for residues. A common problem associated with detergent use is its composition. Many detergent suppliers will not provide specific composition, which makes it difficult for the user to evaluate residues. As with product residues, it is important and it is expected that the manufacturer evaluate the efficiency of the cleaning process for the removal of residues. However, unlike product residues, it is expected that no (or for ultra sensitive analytical test methods—very low) detergent levels remain after cleaning. Detergents are not part of the manufacturing process and are only added to facilitate cleaning during the cleaning process. Thus, they should be easily removable. Otherwise, a different detergent should be selected.

c. Test Until Clean

Examine and evaluate the level of testing and the retest results since testing until clean is a concept utilized by some manufacturers. They test, resample, and retest equipment or systems until an "acceptable" residue level is attained. For the system or equipment with a validated cleaning process, this practice of resampling should not be utilized and is acceptable only in rare cases. Constant retesting and resampling can show that the cleaning process is not validated since these retests actually document the presence of unacceptable residue and contaminants from an ineffective cleaning process.

VII. REFERENCES

1) J. Rodehamel, "Cleaning and Maintenance," Pgs 82–87, University of Wisconsin's Control Procedures in Drug Production Seminar, July 17–22, 1966, William Blockstein, Editor, Published by the University of Wisconsin, L.O.C.#66–64234.

2) J.A. Constance, "Why Some Dust Control Exhaust Systems Don't Work," Pharm. Eng., January-February, 24–26 (1983).

3) S.W. Harder, "The Validation of Cleaning Procedures," Pharm. Technol. 8 (5), 29–34 (1984).

4) W.J. Mead, "Maintenance: Its Interrelationship with Drug Quality," Pharm. Eng. 7(3), 29–33 (1987).

5) J.A. Smith, "A Modified Swabbing Technique for Validation of Detergent Residues in Clean-in-Place Systems," Pharm. Technol. 16(1), 60–66 (1992).

6) Fourman, G.L. and Mullen, M.V., "Determining Cleaning Validation Acceptance Limits for Pharmaceutical Manufacturing Operations," Pharm. Technol. 17(4), 54–60 (1993).

7) McCormick, P.Y. and Cullen, L.F., in Pharmaceutical Process Validation, 2nd Ed., edited by I.R. Berry and R.A. Nash, 319–349 (1993).

Appendix B

Cleaning Validation Glossary

Below is an alphabetical list of terms that may be used in discussions of cleaning validation. All definitions or explanations are to be taken as applied specifically in the context of such cleaning validation.

Acceptance limit: The amount or concentration of target residue above which possible contamination of selected subsequently manufactured products would be rejected. Care must be used in how this term is applied, because it can refer to the acceptable concentration in the next product, the acceptable surface concentration in the manufacturing equipment, the acceptable amount or concentration in the analyzed sample, or the acceptable concentration in a rinse sample. The usual process is to first calculate the acceptance limit in the subsequent product and then to work backward to arrive at the acceptance limit in the selected sample.

Agitated immersion: A system of cleaning in which a manufacturing vessel is flooded with cleaning solution; the cleaning solution is agitated, such as with an agitator, in that vessel.

Agitation: The mixing of the cleaning solution in the cleaning system. Agitation may occur from flow of the cleaning solution through piping or spray devices, or it may be caused by mixers in the equipment. Agitation is generally beneficial in a system because it continually supplies fresh cleaning solution to the surfaces to be cleaned.

Bioburden: The level of microorganisms present in a system or surface. For a sanitizing, disinfecting, or sterilizing process, the higher the bioburden, the more aggressive the antimicrobial process has to be (for example, a longer time and/or higher chemical concentration may be necessary). One way to reduce bioburden is through the cleaning process; this process may either kill or just physically remove microorganisms.

CIP: Clean-in-place, a system or process of cleaning that involves cleaning equipment without disassembly of the equipment. CIP systems usually include a CIP unit [composed of storage tank(s), a heat exchanger, chemical feed equipment, circulation pump, process control devices, and some instrumentation], one or more spray devices, and associated piping.

Cleaning agent: The chemical agent or solution used for cleaning. This sometimes refers to the concentrated cleaning agent and sometimes to the use-dilution of that concentration (such as 5 percent v/v in water). Cleaning agents may be formulated products, commodity chemicals (such as phosphoric acid), or solvents (such as acetone).

Concentrate: The concentrated form of a formulated cleaning agent as sold by the product's manufacturer. This concentrate is usually diluted with water for use.

Contaminant: Something that at a high enough level can or may potentially contaminate an equipment surface, making the subsequently manufactured product unacceptable for use. Contaminants are those items you want to remove to an acceptable level during the cleaning process. Contaminants could include drug active species, degradation products (degraded during the cleaning process), drug excipients, and cleaning agents. Contaminants could also include microorganisms or external contaminants that could render the equipment unacceptable for manufacturing purposes because of contamination levels achieved during storage (and after cleaning).

COP: Clean-out-of place, a system or process of cleaning that involves disassembly of the equipment before cleaning. This disassembly may lead to manual cleaning of the vessel or to cleaning smaller parts in a separate mechanical system, such as a parts washer. Generally, COP is less preferred as compared to clean-in-place; however, clean-in-place cannot be applied to all systems because it usually has to be designed into a system.

Coupon: A small model surface that can be used for either laboratory testing of cleaning performance or for analytical recovery studies for swabbing procedures. For example, a coupon for modeling a stainless steel surface may be a 8 cm × 16 cm piece of 316L electropolished stainless steel.

Cycle development: Work done before the validation protocol to establish a cleaning Standard Operating Procedure. This work may include lab studies on pilot-scale or partial-scale production equipment. The purpose of cycle development is to define a rugged cleaning procedure that will be validatable. Items fixed or established in cycle development include the cleaning agent, its concentration, the cleaning and rinsing time, cleaning temperature, and a myriad of other process parameters.

Dead leg: An area, usually associated with process piping, that leads nowhere (think of it as a dead-end street). Because there is little flow and/or agitation in a dead leg, cleaning may take longer. As a general rule in the pharmaceutical industry, the length of a dead leg should be no more than 1.5 pipe diameters. Orientation of a dead leg can also be important. The ideal is to have a system with no dead legs; in the real world, this is nearly impossible. Dead legs are good candidates for sampling for worst-case cleaning locations in a cleaning validation study.

DI water: Deionized water or water in which all of the ionized species (and, most importantly the hard water salts) have been removed. Hard water salts can interfere with the activity of a cleaning agent, and they can precipitate on equipment surfaces during the cleaning process. DI water can be used for cleaning and preliminary rinsing.

Endotoxin: Toxin that is present in the cell walls of certain gram-negative bacteria. Endotoxins can cause fever and or sickness when injected into the bloodstream (and hence are also known as pyrogens). Endotoxins are generally assayed using the LAL (Limulus amebocyte lysate) testing procedure. Endotoxins are not deactivated by most sanitizing processes or by steam sterilization. They can be deactivated by very high heat or by physical removal during the cleaning process.

Equipment train: A series of individual pieces of equipment linked together for a given process. For clean-in-place of a typical equipment train, the entire train may be cleaned as one system, or the individual pieces can be isolated (by valves, for example) and cleaned separately.

Finish (surface): The degree of roughness or smoothness of a surface. Generally, the smoother the surface, the easier it is to clean.

Grouping strategy: A validation strategy in which products produced on the same equipment and cleaned by the same process are grouped together for cleaning validation purposes. Cleaning validation is performed on the most difficult to clean product, and that successful cleaning validation is considered to apply to all products within the group. Adequate scientific justification is needed to select the most difficult to clean product. An alternative grouping strategy is to group by different equipment sizes (storage tanks of the same design, but of different sizes). Grouping strategies require careful planning and have a higher element of risk. However, for multiproduct manufacturing equipment, they can be a much more efficient way to validate cleaning.

Impingement: The process of a cleaning solution striking a surface. Impingement usually occurs in a spray process. The mechanical action of droplets of cleaning solution striking the surface can help to dislodge and/or to more effectively remove soils from surfaces. Impingement may also help improve agitation, but the effects of impingement and agitation are different.

Interference: Something (other than the target analyte) in an analyzed sample that causes the results of the analysis to be imprecise because that something either adds to or subtracts from the detector response or otherwise interferes with the assay for the target residue. A specific test for a given species takes into consideration possible interferences and is designed so those interfering species are accounted for in some way.

IQ: Installation Qualification, the part of validation that documents that the cleaning equipment is installed according to specifications.

LOD: Limit of detection, the lowest level of an analyte that can be detected but not necessarily quantitated.

LOQ: Limit of quantitation, the lowest level of analyte that can be reliably measured with suitable accuracy and precision.

Master plan: A document that describes how cleaning validation is performed in a given facility. A cleaning validation master plan describes the approaches to establishing limits, cleaning, sampling, and selecting analytical methods. It provides an overview so that there is consistency within the facility. It also covers responsibilities and provides templates for cleaning procedures and protocols. A master plan is not required, but it helps significantly.

Monitoring: The process of evaluating the cleaning process on a routine basis during and after the cleaning process. It may involve visual inspection as well as sensors that record parameters such as temperatures, pressures, and conductivities. The tests that are done during monitoring of cleaning are not necessarily the same tests that are done during cleaning validation. Monitoring serves as a control that can identify trends before action limits are achieved.

Neutralization: The process of modifying the pH of a used alkaline or acidic aqueous cleaning solution to render it more suitable for discharge into a waste treatment system. Most typically, the pH needs to be adjusted to the 5.5–9.5 region. As a general rule, pH neutralization of a cleaning agent solution should not be done in the process vessel that has been cleaned; such a pH change may cause the cleaned product to redeposit within the system. Neutralization is best done in a separate vessel or in an in-line neutralization system as the spent cleaning solution is drained from the process equipment.

Once-through: A clean-in-place process in which the cleaning solution passes through the spray device, through the equipment to be cleaned, and goes directly to the drain. This is not commonly done for the cleaning solution because of the expense. As applied to the rinsing step in a clean-in-place system, once-through covers rinsing systems in which the rinse water goes directly to drain after passing through the equipment to be rinsed. Once-through is commonly used for the rinsing process because it minimizes the possibility of recontamination of rinsed surfaces. Once-through systems are to be contrasted with "recirculation" systems.

OQ: Operational Qualification, that part of the validation that documents that the cleaning equipment operates correctly. This might include documentation of such items as the spray pattern in a clean-in-place system and pump flow rates.

Organoleptic: Measured by the senses. Visual examination is a type of organoleptic procedure used in cleaning validation studies.

Pitch: The slope of pipes in the cleaning system. The slope should be to the drain, because the purpose of the slope is to minimize residual water in the system after it is cleaned. The typical target pitch is 1/16 inch per foot of length of pipe (about 5 millimeters per meter).

Placebo sampling: A method of sampling that uses the manufacture of a product placebo to sample the cleaned equipment. The placebo is analyzed for the presence of the target residue. This is not a sampling method preferred by the Food and Drug Administration, because of concerns about nonuniform contamination and the analytical power of a method to measure the target residue in the placebo matrix.

PQ: Process Qualification, that part of the validation that documents that in three consecutive cleaning trials, the cleaning process consistently produces equipment cleaned to the preestablished residue limits.

Prevalidation: Work done before one actually approves and executes the validation protocol. This may include cycle development work on the Standard Operating Procedure, establishment of residue limits, analytical method selection and validation, sampling method selection, recovery verification, and grouping strategy studies.

PW: USP Purified Water, specially manufactured and controlled water with low conductivity and with low total organic carbon. See the *United States Pharmacopeia* for more detail.

Pyrogen: *See* **Endotoxin.**

Recirculation: A clean-in-place process in which the cleaning solution passes through a spray device, through the equipment to be cleaned, and returns to the cleaning solution tank. The cleaning solution is then circulated through the spray device and equipment train again. This is commonly done for the cleaning step but not for the rinsing step. Recirculation systems are to be contrasted with "once-through" systems.

Recovery: For analytical procedures using swab or rinse sampling, the percent of the known amount of target residue spiked onto a coupon that is actually measured in the analytical procedure. Ideally, recoveries should be in the 80–120 percent range. The amount spiked should correspond to the amount found near the acceptance limit for that residue.

Residue: Whatever is left behind after the cleaning process. Residues may include drug actives, drug product excipients, process aids, microorganisms, various degradation products, and cleaning agents. Residues are measurement targets in cleaning validation protocols to see if acceptance limits are met. *See also* **contaminant.**

Revalidation: A regular (for example, every two years) process by which the performance of a validated cleaning process is evaluated to determine whether the cleaning process is still validated. This might include a review of monitoring data, cleaning logs, and any changes to the system done under a change control process. This review and the judgment as to whether the process is still under validation should be documented. Often, revalidation is used to refer to the process of validating a process after a major change. However, in a technical sense, one is not *re*validating a system; one is validating what is (for validation purposes) a *new* process. The two uses of the term really apply to different situations. The process for revalidation of a cleaning process should be defined in the cleaning validation master plan.

Riboflavin testing: A procedure for testing the spray pattern from a spray device by coating the interior equipment surfaces with a dilute solution of riboflavin, running water through the system for a short time, then opening up the system and examining interior surfaces with a black light. Riboflavin will fluoresce under a black light, so areas that the spray is inadequately contacting can be identified. This test is sometimes used for determining worst-case cleaning locations, which can be a misuse of this test.

Rinse sampling: A procedure for sampling surfaces that involves flooding the surfaces with rinse water (or another solvent) to effectively remove target residues from the surfaces. The sampling rinse is then analyzed for the target residue. Recovery of the target residue by the rinse sampling procedure must be demonstrated.

Sensitivity: This is commonly confused with the limit of quantitation or the limit of detection. However, for analytical chemist purists, sensitivity is really the "slope of the working curve" and has to do with the relative change in detector units as a function of change in analyte concentration. In general, a more sensitive method rather than a less sensitive method is desired.

Shadow area: In a clean-in-place process, any area that does not adequately receive spray from the spray device because of some impediment within the process vessel. For example, in a tank with a top agitator and only one offset spray ball, there will be an area on the opposite wall that will not be adequately contacted with cleaning solution because it is blocked by the agitator shaft. Any shadow area can be detected by riboflavin testing. If shadow areas exist, it is necessary to change the orientation or number of spray devices to make sure that all surfaces are adequately contacted with the cleaning solution.

Shared product surface area: The surface area within the process equipment that has the potential to directly transfer a contaminant to the subsequently manufactured product. These surfaces should be focused on during the cleaning process. When the acceptance limits per surface area of equipment are calculated, the shared product surface area is required.

Soil: The material in process equipment to be removed during the cleaning process. This includes the drug active, drug product excipients, and process aids (such as fermentation broth or process solvents).

SOP: Standard operating procedure, a written document detailing the specific steps to be taken in the cleaning process. In some facilities, it might be called a work instruction rather than a SOP. The FDA will want to see cleaning SOPs if they take a look at cleaning validation; cleaning validation can only be done once a cleaning SOP is defined.

Specificity: The ability of an analytical method to unequivocally assess the analyte (target residue) in the presence of components that might also be expected to be present. Specific methods are preferred but not required for cleaning validation studies.

Swab: This is an expensive Q-tip™. It usually consists of cotton on a wooden handle or knit polyester on a plastic handle. A swab is used to effectively remove target residues from the surfaces in swab sampling. Swabs must be carefully selected because they can contribute interferences to analytical procedures.

Swab sampling: A procedure for sampling surfaces involving wiping the surfaces with a swab (usually a swab wetted with water or another solvent) to effectively remove target residues from the surfaces. For chemical analysis, the swab is then desorbed into a greater quantity of water or solvent and analyzed for the target residue. Recovery of the target residue by the swab sampling procedure must be demonstrated. For microbiological sampling, the swab is desorbed into a buffer solution, and a plate count is done.

TOC: Total organic carbon, a nonspecific analytical procedure that can be used in cleaning validation studies.

Use-dilution: A relatively dilute solution of a cleaning agent concentrate. This is the concentration of the cleaning agent as it is prepared for use. For example, most concentrates are commonly used as cleaning agents at a use-dilution from 1 to 10 percent by volume in water.

Validation: Documented evidence with a high degree of certainty that a cleaning process will consistently produce product contact surfaces meeting their preestablished residue acceptance limits.

Verification: Documented evidence that an individual specific cleaning event has produced product contact surfaces that are acceptably clean. Verification should be contrasted with validation. While validation involves a repeating process (where one can conduct at least three process qualification runs), verification involves testing a specific cleaning event. Data obtained in a verification study, while suggestive of what might happen in the future if the process is repeated, should only be used to support that one specific cleaning event it is associated with.

WFI: Water for injection, essentially Purified Water that is endotoxin free and with a lower level of microbes.

Worst case: The FDA requires the evaluation of cleaning Standard Operating Procedures under worst-case conditions. As applied to process conditions, worst cases mean those conditions within normal operating parameters most likely to cause process failure. For example, if the cleaning temperature is controlled from 55 to 65°C, then the worst-case cleaning might be at the lower end of the temperature or around 55°C. As applied to sampling locations, worst case means those locations in the equipment most likely to have the highest levels of residue after cleaning (these locations may be the most difficult-to-clean locations). As applied to recoveries, worst case means those recovery conditions (within normal recovery procedures) giving the most likely chance that limits are exceeded. For example, if swab recoveries are established as 80 percent for $1 \ \mu g/cm^2$, 85 percent for $5 \ \mu g/cm^2$, and 90 percent for $10 \ \mu g/cm^2$, the 80 percent recovery value should be used for cleaning analyses. As applied to acceptance limit calculations for contamination of a subsequent product, worst case means calculating the minimum dose of the target active residue in the maximum daily dose of the subsequently manufactured product. As applied to product grouping strategies, worst case mean selecting the representative product for testing in the three validation process qualifications runs as that product that has been demonstrated to be most difficult to clean.

Index

Printed in the United States
by Baker & Taylor Publisher Services